VINE VARIETIES CLONES AND ROOTSTOCKS FOR VINEYARDS IN GREAT BRITAIN

and other cool climates

3rd edition – 2026

VINE VARIETIES CLONES AND ROOTSTOCKS FOR VINEYARDS IN GREAT BRITAIN and other cool climates

A guide to the varieties of grape vine, clones and rootstocks suitable for wine production in Great Britain and other cool climates

STEPHEN SKELTON MW

Published by the Author
January 2026

S. P. Skelton Limited,
22 Radipole Road,
London,
SW6 5DL.
E-mail: mail@stephenskelton.com
Telephone: 07768 583700

First published in paperback in Great Britain in 2014. Second Edition published in paperback in Great Britain in 2020. Third edition published in paperback in Great Britain in 2026.
Edition: 3-001-c

A CIP catalogue record for this book is available from the British Library.

ISBN: 978-1-9163296-5-2

Books written by Stephen Skelton MW
"Viticulture – An introduction to commercial grape growing for wine production" 2nd edition 2020. Also published in Japanese and Chinese.
"The UK Vineyards Guide 2016"
"Wine Growing in Great Britain 2nd Edition – a complete guide to growing grapes for wine production in cool climates" 2020
"Vine varieties, clones and rootstocks 2nd Edition – A guide to the varieties of grape vine, clones and rootstocks suitable for wine production in Great Britain and other cool climates" 2020. 3rd edition published in 2026
"The Wines of Great Britain" (Published by Infinite Ideas Classic Wine Library) 2019
"The Knight who Invented Champagne" 2021

In memoriam
Peter Hall 1943 - 2025

This book is dedicated to the memory of my great friend and colleague Peter Hall who planted his vineyard at Breaky Bottom in East Sussex in 1974 and ran it and made wine there for over 50 years. Here is a picture of him standing in front of his beloved Seyval blanc vines. We may have disagreed about a few things, but never the qualities of Seyval blanc.
Sleep well my friend.

Contents

Introduction 1

1. Choosing the right variety of vine 3
- Which varieties should I plant? 4
- Varietal change 6
- Still or sparkling? 8
- The change from still to sparkling 8
- Still wines fight back 9
- Rise of the Crouch Valley winegrowing region 10
- Oversupply of English and Welsh sparkling wine? 12
- Sales of English and Welsh wines 13
- Sales by sector 13
- Stocks of English & Welsh wine 14
- Where will it all sell? 15
- Still or sparkling? A summary 15

2. Recommended vine varieties for Great Britain 17
- Still white wines 17
- Still white PIWIs and hybrids 20
- Varieties NOT recommended 20
- Still red wines (including red PIWIs) 20
- Still rosé wines 22
- White traditional-method sparkling wines 22
- Rosé and red traditional-method sparkling wines 23
- Novel styles of still & sparkling wines 23
- Vine varieties for less favourable regions and cooler sites 25

3. Performance of vine varieties 28
- Yields in Great Britain 28
- Performance of PIWIs 29
- Yields in regions 29
- Potential alcohol levels 31
- Acidity levels 32

4. Vine varieties – the legal situation 34
- Categories of wine that may be produced in England and Wales 35

5. Vine variety descriptions – *Vinifera* varieties 37

Acolon	38	Meunier	48
Albariño	39	Müller-Thurgau	49
Alzey 7672*	39	Ortega	51
Auxerrois	41	Pinot blanc	51
Bacchus	41	Pinot gris	52
Chardonnay	42	Pinot noir	53
Dornfelder	44	Pinot noir Précoce	55
Frühburgunder, Blauer	45	Reichensteiner	55
Gamay noir	46	Sauvignon blanc	56
Huxelrebe	47	Schönburger	57
Madeleine x Angevine 7672	47	Siegerrebe	58

* Alzey 7672 is now the correct name for Madeleine x Angevine 7672

6. Clones of Chardonnay, Pinot noir and Meunier 60

7. PIWIs – Fungus resistant vine varieties 65

Introduction 65
Some history 66
Early hybrids and interspecific crosses 67
Modern interspecific crosses and disease-resistant hybrid crosses (PIWIs) 68

8. Selecting PIWI and hybrid varieties to plant in Great Britain 71

Are the most widely planted PIWI varieties the best? 71
Most widely planted PIWIs and hybrids in Europe 2025 71
Germany 71
Austria 73
Switzerland 74
Belgium 75
France 76
Netherlands 76
Minor European winegrowing countries 76
PIWIs in New Zealand 77
Do PIWIs produce as good a wine as vinifera varieties? 78
PIWI International Wine Competition 78
To summarise: Do PIWIs produce as good a wine as vinifera varieties? 79
Genetically modified vines 81

9. PIWI and hybrid vine varieties for Great Britain 83

Current situation of PIWIs and hybrids in GB 83
Recommended PIWI and hybrid vine varieties for vineyards in Great Britain 85
Descriptions of 31 PIWI and hybrid varieties 87

10. Rootstocks for Great Britain **105**

Appendices

I. The Complete Story of Sparkling Wine production in Great Britain 1662-2024 109

II. The story of Wrotham Pinot 120

III. Distribution of vineyards in Great Britain 123

IV. Wye College Experimental Vineyard 124

Introduction

3rd Edition 2026

This, the 3rd edition of *Vine Varieties, Clones and Rootstocks,* comes at a time of change in vinegrowing and winemaking in Great Britain. Change for several reasons: the first is a combination of Brexit, the war in Ukraine, and the cost-of-living crisis that have all increased production costs on the supply side, and reduced people's incomes on the demand side. Both of these have had an impact upon sales of wine. The second is the fact that the planted area of vines in GB has continued to rise, seemingly unaffected by external forces. Since the last edition of this book was published in 2020, the area under vine has increased from around 3,500-ha (8,650-acres) to at least 5,000-ha (11,490-acres) and probably more, a rise of almost 43 per cent.[1] This of course means that production levels have also risen. The average number of 75 cl bottles produced per year in the years 2015-21 was 7.98 mn; between 2022 and 2024 it was 14.82 mn bottles, almost twice as much. If we add in to those figures the above-average 2025 yield, then the total will be nearer 16 mn bottles a year.

This rise in production is fuelled not only by the increase in vineyard area, by also by rising yields, the result of better site selection, better vine husbandry and of course, climate change. In the 13 years between 2005 and 2017, the official yields averaged 21.38 hl-ha a year. Between 2018 and 2024 they were 32.08 hl-ha, an increase of 48 per cent. Given that the area in production in 3 years' time will be nearer to 5,000-ha (12,355-acres), and maybe there will have been a further rise in per hectare production levels, we can then expect annual production to be averaging around 20 mn bottles.

Sales, however, have not kept pace with either plantings or availability, and stocks are rising. It is estimated that there are now around 40-50 mn bottles of sparkling wine in producer's stores which is six to seven years' worth of annual sales. What these figures mean is open to conjecture, but some certainly think that the industry is due for some consolidation and for many producers, stock reduction will be towards the top of their 'to do' lists.

The third reason for change, at least as far as the subject of this book is concerned, is perhaps a more positive one. Five to ten years ago, the words 'sustainability' and 'regenerative' barely registered with farmers and growers, let alone the public. Today they are becoming much better known and the numbers of farmers and growers practicing organic and biodynamic agriculture and horticulture in GB is growing. Some of the impetus for this (or 'blame' depending on your point of view) goes back to the European Union who in 2009 proposed a regulation called 'The Sustainable Use of Plant Protection Products' which included EU wide targets to reduce the use of chemical pesticides by 50 per cent by 2030. Naturally this caused shockwaves of protests by agrochemical companies and industrial farming practitioners who claimed that food production would 'fall by 50 per cent' and prices would rise out of control. However, for vines, a crop which per hectare is a well-known disproportionate user of pesticides[2] (although on a crop value basis

1 The official figures from the Wine Standards (WS) team at the Food Standards Agency, issued annually along with the previous year's production figures, tend to lag behind reality as the requirement to report new plantings within 6 months of planting is not enforced. In addition, WS keep two databases, one for planted area and another for grape varieties which never match up. The latest (2024) vineyard area stands at 4,841-ha of all vineyards, of which 4,745-ha are commercial, and 3,763-ha were in production in 2024. The latest vine variety database records 4,651-ha of vineyards of all types.

2 By 'pesticides' I mean all plant protection materials applied to the vines, including those allowed under organic and biodynamic regulations, whether by spraying, dusting or soil applied. The term also included herbicides (weedkillers).

probably not quite so greedy), what this proposal also did was to put in the spotlight vine varieties with inbuilt natural resistance, ones we now call PIWIs.[3]

As you will discover as you read this book, hybridising vine varieties has been practiced since the 1840s and probably much earlier. However, when the twin horrors of first Powdery and then Downy Mildew arrived from the fields and forests of the eastern United States, plant breeders quickly realised that some American vine species had disease-resisting properties that the European species did not and that by cross-breeding the two, they could produce vines that stayed healthy and whose grapes could also make palatable wine. The arrival of the far more dangerous and life-changing insect *Phylloxera*[4] in 1863 turbocharged the vine breeding world with the distinct promise that cross breeding could produce not only disease-resistant vines, but also insect-resistant vines. And even more importantly, vines that did not need grafting and which could be grown directly in the place where they would stay for the rest of their lives, the so-called HPDs – *hybrides producteurs directs*. Well, as it happened, things didn't quite work out like this and after many decades of trying, HPDs were banned from the production of quality wine and growers were much happier to carry on with their old tried and tested varieties, grafted on to *Phylloxera* tolerant rootstock and helped along by better and better plant protection products. But vine breeders never gave up hope and as you will find out in the chapter on PIWIs and hybrids, some regions still use them and value them and their numbers are rising. This book is an attempt to summarise where we are with the new PIWI varieties and to provide some research-based recommendations and suggestions for growers in Great Britain, and other cool climate regions. I hope you find the book helpful.

Stephen Skelton MW
Fulham 2026

3 PIWI stands for *Pilzwiderstandsfähig* (meaning 'fungus resistant' in German) and is used as a shorthand to describe new crosses of vines mostly made since 1945 having both *vinifera* and non-*vinifera* parentage.

4 Initially named *Phylloxera vastatrix* (the destroyer), it is now officially called *Daktulosphaira vitifoliae.* It was called *Phylloxera* because of its similarity to the insect that is responsible for oak-leaf galls *Phylloxera quercus.*

Chapter 1
Choosing the right variety of vine

The vineyard at Painshill Place, first planted in 1740 and replanted in 1989

When in 1975, I started thinking about planting a vineyard in Britain, the options for new varieties were fairly limited. Although I stuck with Müller-Thurgau and Seyval blanc as my two main commercial varieties (along with most of the other growers at the time), I also planted 1,250 Reichensteiner vines and the same number of Gutenborner, a variety only being grown then by one other British grower, Carr Taylor. Gutenborner was recommended to me by Geisenheim's Professor Kiefer who had been helping with the establishment of the nearby vineyards at Lamberhurst for Kenneth McAlpine and so had some experience of growing conditions in Britain[5]. These four varieties made up the bulk of the 2-ha (5-acres) I initially planted. My almost two years in Germany, working in a vineyard and winery and studying at Geisenheim had not only been spent in learning how to grow grapes and the rudiments of winemaking, but I had also spent a considerable amount of time researching into vine varieties in an attempt to discover if there were any others that were then unplanted in Britain, but which might be suitable. I remember very well going to see Professor Becker, the late and much-lamented head of the vine breeding department, with a list of thirteen varieties I thought of as 'possibles' for Britain, to ask his advice. He almost grabbed the list and said, 'Come back on Thursday,' which I took to be a signal that he was too busy to discuss the topic but might have time later in the week. I duly turned up on Thursday to find that his assistant, Ernst Abel (who had been over to Lamberhurst to help make their maiden vintage in 1976), had been instructed to get a bottle of wine from each of the varieties that I had listed (the research cellar made trial batches of wine from hundreds of varieties and clones each year) and have them lined up in the cellar ready for me to taste!

The result of this was that I planted, in addition to the four main varieties mentioned at the start of this chapter, one hundred vines each of Ortega, Schönburger, and Rabaner, plus twenty-five vines each of Ehrenfelser, Huxelrebe, Kerner, Optima, Siegerrebe and a Riesling x Silvaner cross called Gm 4-46. One of the varieties I had on my list and tried at the tasting was Bacchus and I noted on my tasting notes (which I still have, dated 15 Nov 1976) 'Acid too high. Wine not good if under 75°OE' and was advised not to plant it! In my trial plot I also planted a range of other varieties: Auxerrois, Black Hamburgh, Chasselas Rosa, Chenin blanc, Dunkelfelder, Gewürztraminer, Gold Riesling, Léon Millot, Meunier, Pinot blanc, Pinot gris, and Pinot noir. It was an ambitious list, but I thought that I would try to create a mini-research station of my own. Today, if faced with the same task – to select some new, untried varieties for planting in Britain – it would be considerably harder to choose a reasonable number, given the huge range of different varieties now available.

5 The Geisenheim Institute for Grapevine Breeding is arguably the pre-eminent research station for developing new vine varieties and improving strains of older ones through clonal selection. It is situated in the centre of Germany's best winegrowing region, the Rheingau.

In order to suggest new varieties for Britain, we have to get over the problems which all new variety recommendations suffer from: none of them has been bred specifically for British climatic conditions and none rigorously tested in Britain. We have therefore to sift through the claims made for each variety and decide for ourselves how they relate to British problems. Disease resistance is obviously a desirable trait but should it be the deciding factor? If you had a choice between a variety that cropped well, produced ripe grapes with moderate acidity that made good wine, yet required a full spray programme, typically 12-14 sprays, compared to say a variety that cropped well, needed say 3-4 sprays a season, but whose wine was of a lower quality, which would you choose? For me, the first attribute for a variety to be preferred over the ones we currently grow is the quality of its wine. A reasonable yield is very welcome, disease resistance is admirable, but without wine of sufficient quality to lift it over the problems associated with introducing a new variety to the wine trade and the wine-buying public, the variety is going to struggle commercially. This 'quality or resistance' conundrum is one only you can answer.

This problem is not a new one. Varieties such as Faberrebe, Findling, Kerner, Kernling, Optima and Würzer have all had their day in the British spotlight, were found lacking and have mainly been grubbed up. Newer varieties that have found favour in vineyards in Britain and are expanding – Acolon, Orion, Phoenix, Regent, Rondo, and Solaris – are, with the exception of Acolon, all modern interspecific PIWI crosses. The reds, Regent and Rondo, have been chosen because they are relatively disease resistant and produce wines with plenty of colour. If you are hell-bent on producing a deep red wine in a very cool climate, then these are much better in terms of wine quality than the old red hybrids we used to grow: Cascade (Seibel 13/053), Léon Millot, Maréchal Foch and Triomphe. The others, all whites, have mainly been planted in vineyards in the more climatically challenged areas of Britain, again because of their resistance to disease and their suitability for cooler climates.

Others creeping up the charts are either standard varieties such as Pinot blanc, Pinot gris, and Sauvignon blanc, which have shown that in good years and on good sites they can make acceptable (sometimes even very good) wines, or newcomers such as Cabinet Cortis, Cabaret noir, and Divico, all new PIWI varieties. At the bottom of the list of currently planted vines are around 45 varieties, most of which have appeared since 2018 when the last full vine variety survey was carried out, all of which are planted at less than 1.00-ha. Some of these must be considered as purely experimental and whether or not they will succeed is uncertain – only time will tell. Others such as Chenin blanc, Grenache noir, Cabernet Sauvignon, and Nebbiolo seem decidedly risky, but again, who knows? Bacchus after all was once an untried variety.

Which varieties should I plant?

After you have chosen your site, the next major decision you face is which varieties to plant? The answer, strangely, is actually quite easy. They should be the ones that (a) suit your site and (b) make the style of wine you will be able to sell! Of course, this is far easier said than done. Given that on a site with no previous history of growing vines – a common situation for many winegrowers in Britain – prior knowledge of what varieties 'suit' your site is, of course, difficult, but not impossible. And again, the question of what style of wine 'you will be able to sell' may be something you have not considered and is therefore also quite difficult to answer.

However, asking these two questions brings to the fore two essential points: the quality of your site will dictate the best varieties for you to establish on it, and your need to have a vision of the wine you would like to produce and how and where you propose to sell it before you plant, and will hep you select suitable varieties for that wine. Do not fall into the trap of growing a variety that your site will not support, or attempting to make a wine for which there is a limited market, at least not before considering the pros and cons. I often ask potential growers at my first meeting, 'Which English wines have you enjoyed recently?' or, 'Tell me what styles and types of English wines you like?' and it always amazes me how many look rather blank, shuffle around a bit and then mention one or two wines that they vaguely recall having tasted. In order to have a successful vineyard, especially if you are proposing to make and sell the wine, it is imperative that you should know and enjoy the product.

The range of varieties available to plant is huge and, to the uninitiated, the question of which to plant is either a problem to be run away from or an opportunity to experiment. There are advantages and disadvantages to both views. For those who see the selection of varieties as a problem, the answer is to stick to the varieties that have proved themselves and which suit your site and the style of wine you wish to make. Do not try to re-invent the wheel. If you were a grower in Chablis or Sancerre, Burgundy or Alsace would you be worried about what variety to plant? No. The *appellation* has already determined which varieties suit your region and while you may have the choice of clone – although even this is not certain – your task is to grow the best grapes you can. There is a reason why the four most widely planted varieties in GB – Chardonnay, Meunier, Pinot noir and Bacchus – account for 78 per cent of all vineyards and if you take the south-east corner of GB (Essex, Kent, East and West Sussex, Hampshire and Surrey) which are home to 74 per cent of all GB vineyards, the percentage of these four varieties rises to over 90 per cent of all the vineyards planted in this region. They are the varieties that have a proven track record in both yield and quality, and they are the varieties that the wine buying public recognise.

Apart from site selection, the choice of the varieties that suit your site, your financial position and your marketing plans is the most important decision you will have to take in the establishment of your vineyard. In taking this decision, the best advice I can give is to learn from others and to look backwards to what has been successful, rather than to look forwards and hope the obscure varieties you have heard about will produce award-winning wines. Or at least, do not plant too many of them and then be prepared for both success and failure.

When looking backwards, I am guided by the collective wisdom of the 1,000-plus growers in Britain, by the membership of the EVA, the UKVA and WineGB[6] and of the judges who judge the wines entered into the annual 'Wine of the Year' competition. If you just take the trophies and gold medals for both still and sparkling, and white, red and rosé wines, the list of successful varieties is actually quite small. The only real criterion is to choose those varieties that have proved themselves capable of making high-quality wines, because it is only by making wines of high quality that your enterprise will be successful in the long run. I accept that looking back has its problems, as growers can only make wines from the varieties they have in the ground and, therefore, some of

6 The EVA (English Vineyards Association) and the UKVA (United Kingdom Vineyards Association) were earlier incarnations of what today is called Wines of Great Britain (WineGB).

today's (and yesterday's) wines are made from varieties that, given a fresh start, many growers would never plant today and will certainly not plant again. Please also remember that GB is not a region where high average yields are possible – that is, high in world terms – thereby hoping to make wine that will compete on price with wines produced in regions where high yields occur regularly. This is not possible if you want to create a (financially) sustainable enterprise.

Varietal change

For those who see the selection of a variety as an opportunity to experiment, let me issue a warning straightaway. Although we have seen considerable changes in the spectrum of varieties being grown in Britain since modern commercial vineyards started in the 1950s, some of the old favourites are still there. The top ten varieties following the first compulsory survey in 1989 when the total planted area was shown to be 928.51-ha (2,294-acres) were (in order of area planted): Müller-Thurgau, Reichensteiner, Seyval blanc, Bacchus, Schönburger, Huxelrebe, Madeleine x Angevine 7672, Pinot noir, Kerner and Ortega. Thirteen years later, in 2002, when the area under vine had shrunk to 812-ha (2,006-acres) the top eight varieties were, not surprisingly, almost exactly the same! Only Kerner had dropped out (to number 12) to be replaced by Chardonnay, creeping in at number 9. However, jump another twenty-two years to 2024, when the total area under vine has risen by almost six times, and the variety spectrum in the top ten looks a little different. The old favourites Bacchus, Reichensteiner and Seyval blanc are still there, the last two hanging on by their tendrils, but the list is, of course, dominated by Chardonnay and Pinot noir, which together occupy 62 per cent of the planted area. The other 'old school' varieties have all dropped out of the top ten: Ortega to number 12, Müller-Thurgau to number 13, Madeleine x Angevine 7672 (now officially renamed as Alzey 7672) to number 15 and Huxelrebe to number 21. They have been replaced by Meunier, Blauer Frühburgunder (Pinot noir Précoce), Pinot gris, Solaris, and Rondo – two red varieties, one white and two PIWIs. See Table 1.

Major Vine Varieties in GB 2024

No.	Variety	Ha	%
1	Chardonnay	1498.28	32.2%
2	Pinot noir	1369.77	29.5%
3	Meunier	410.78	8.8%
4	Bacchus	333.81	7.2%
5	Seyval blanc	122.71	2.6%
6	Solaris	117.07	2.5%
7	Frühburgunder, Blauer	95.00	2.0%
8	Pinot gris (Ruländer)	79.94	1.7%
9	Reichensteiner	64.53	1.4%
10	Rondo	64.23	1.4%
11	Sauvignon blanc	56.17	1.2%
12	Pinot blanc	53.85	1.2%
13	Ortega	45.17	1.0%
14	Müller-Thurgau	37.43	0.8%
15	Alzey 7672	36.62	0.8%
16	Regent	32.42	0.7%
17	Phoenix	28.95	0.6%
18	Schönburger	19.92	0.4%
19	Siegerrebe	18.34	0.4%
20	Dornfelder	15.85	0.3%
	Other varieties	150.06	3.2%
	Total hectares	4650.90	100.0%

Source: Wine Team at the Food Standards Agency

Table 1

Not only has the change in varieties over the last thirty years affected the type and style of wines produced in Britain, with Champagne-variety sparkling wine now accounting for around 70 per cent of total production, but it has also changed consumers' perception of our wines. As I have often said, in earlier times, when a wine retailer or sommelier was asked what an English wine was like, he or she might have said, 'Well, it's made from Müller-Thurgau and Schönburger with a touch of Reichensteiner and it tastes a bit like German wine or maybe a wine from Alsace,' and in doing so lose the customer's attention and interest before the second umlaut. Today, when faced with the same question about a bottle of English or Welsh Sparkling Wine, the same retailer or sommelier might say, 'Well, it's just like Champagne,' and then perhaps as an afterthought, in order to make sure the customer was completely at ease, add, 'And it's made from Chardonnay and Pinot noir'. In saying this, in such simple terms, a message which every wine drinker understands has been easily and simply conveyed. I believe this has made a huge difference to the industry and has given the wine trade in all its many guises the confidence to support and to sell, the home-grown product. Remember that wine is one of the very few agriculturally-based consumer products where the name of the variety of plant which the product is made from is on the label, and they will often, even always, select their purchases on this basis. Try asking someone at the checkout what variety of wheat is in their loaf, barley in their beer or potato in their oven-chips. They will have no idea. In wine, the varietal name is very often an important part of the marketing package.

Those vineyards planted with German varieties in the 1970s, '80s and '90s were not defeated by poor quality wine, and they were not necessarily defeated by poor yields. What defeated them was the problem of marketing their (mainly still) wines in the time frame you have to sell still wine. Most new vineyards start cropping in their third or fourth year and, if you are producing still wine, six months after picking, your wines are bottled and ready to sell. Initially, with a small maiden vintage, and with some local interest in your wines, they may sell quite quickly, but come October and another vintage, this time hopefully a full-sized one, is ready for picking and in a few months' time, you have another vintage for sale. The pressure is on.

Buyers, especially trade buyers who perceive most of our still wines as being fresh, fruity, lively wines, at their best within twelve to eighteen months after picking, are not so interested in older vintages and would rather taste your more recent wines. If sales are not matching output, you cannot afford to store your wines in tank indefinitely – there is another harvest approaching – and the temptation (and the reality) is to cut your losses, lower the price and move the stock. In 1984, Kenneth McAlpine, then the owner of Lamberhurst Vineyards, the largest and (seemingly[7]) most successful vineyard in the country, found himself, after the larger than average crops in 1982 and 1983, with one million bottles in his cellars and sales lagging behind production. At the time, the vineyard's best-selling 'Lamberhurst White' sold for £2.99. Faced with the prospect of another large crop in 1984, McAlpine looked at the alternatives: take on more sales staff, advertise, promote, attend more shows and fairs and maintain the price; or slash the price and stimulate sales. He took the latter route, cut the price to £1.99 and within six months had cleared the excess stock. Short term, the problem was cured; long term, it set the benchmark price for entry-level English wine at a low – and definitely unprofitable – point from which it would struggle to recover. Other large-scale producers, such as Carr Taylor (who would face bankruptcy a few years later), Barkham Manor, Penshurst, Three Choirs and Wellow all found it hard to lift their prices above those of their main competitor – Lamberhurst[8]. Sparkling wines, however, are different.

7 I say 'seemingly' as in fact Lamberhurst Vineyards at the time was loss-making, something that didn't seem to trouble Kenneth McAlpine too much as he was able to write it off against other income. When I took over as winemaker and general manager in 1988, I was given audited accounts which showed annual losses running at around £350,000 pa, which equates to over £1.25 mn in 2025 prices, and this level of loss had been going on for several years.

8 At the time I owned Tenterden Vineyards (today's Chapel Down) which was only a few miles from Lamberhurst, and I found that this price reduction didn't have quite the effect I feared, mainly because we hand-sold most of our output from the farm-gate. When people said 'why is your wine more expensive than Lamberhurst's?' We always said: 'Our wines are better and of course you get what you pay for!'

Still or sparkling?

You cannot get into the (bottle-fermented) sparkling wine business without putting the wine into a bottle, putting that bottle into a store (usually in a storage or riddling crate) and stacking the crates in an insulated, secure, temperature-controlled store. Once the wine is in the store, all you can do is wait for it to mature. Therefore, why sell it (a) before the wine is ready and (b) until you can find someone willing to pay what you perceive the right price to be? With sparkling wines, bottle age is seen as a plus: two years, good, four years, better, six years, even better. This puts the decision of when to sell firmly back in the producer's hands and although building up stocks creates havoc with the cash-flow, at least it means that the temptation to cut and run towards an easy sale at a low price is lessened. You can probably sell enough to keep the cash-flow positive, leaving you time to re-double your sales efforts, open that farm-gate outlet, build that website, visit more hotels and restaurants, send out more samples to merchants, attend more shows, investigate export opportunities. In short – find different ways of getting your wine to market. I believe this difference between still and sparkling wines has helped the prices of British-grown sparkling wines to stay resiliently high even though both the number of different sparkling wines and the amount available for sale have risen very considerably in recent years. Whether this price premium for our sparkling wines will remain, in view of the volumes now coming to market, is an interesting question and one which only time will answer.

However, the sparkling wine business requires large amounts of capital to be tied up in stock. In my book, *Wine Growing in Great Britain,* I outline the costs of sparkling wine production and give a calculation of the amount of money tied up in stock (*Chapter 2: Why plant a vineyard in Great Britain?* in the section on *the costs of producing wine*). Taking 2025 prices of growing grapes and wine production, the figures look fairly frightening. If you have a modest crop of say 7.5 tonnes-ha (3 tonnes-acre) and wine production costs of £4.00 a bottle for still wine and £7.00 a bottle for sparkling wine, then per hectare per vintage, you are looking at £28,500 for still wines, and a whopping £39,375 for sparkling. Multiply those by average stock levels of say 3 years for still wines (from bottling to sale) and 5 years for sparkling (from bottling to storage, excluding riddling, disgorging and dosage) and it's a total of £85,500 per hectare (£34,601-acre) for still wines and £196,875-ha (£79,674-acre) for sparkling. The larger vintages of 2018 and 2023 have pushed up stock levels and many producers will have more than this level of stocks. Chapel Down are currently (2025) selling wines from the 2016 vintage, as are Gusbourne and Nyetimber who have wines going back to 2010 still on sale, plus wines from 2013 and 2016. According to Companies House, at year end 2023, Chapel Down had inventory stocks worth £22.58 mn, Gusbourne £15.55 mn, and Nyetimber £41.47 mn. Sparkling wine is a cash-hungry business. Your selection of varieties and wine styles must therefore be taken not only from a marketing perspective, but also from a financial one.

The change from still to sparkling

NEWS FOOD & HEALTH

English sparkling wine beats Champagne

Moët & Chandon and Veuve Clicquot beaten for taste by Sussex-born bubbly

Two English sparkling wines available from M&S, Waitrose and online have beaten Moët & Chandon and Veuve Clicquot Champagnes in a Which? taste test.

The £26 M&S Marksman English sparkling wine impressed our experts in our blind taste test of five bottles of English sparkling wine and 12 bottles of Champagne. They described it as quite crisp and dry with a nice complexity, apple flavours and a creamy texture. Although it is made using the same process as Champagne, it is about £7 cheaper and scored higher than some branded Champagnes.

From the same wine estate, the £24.99 Ridgeview Bloomsbury 2010 makes a great aperitif. Our experts found it pleasant, elegant and easy to drink, with a refreshing lemony finish.

It seems that you can now proudly bring a bottle of English sparkling to the table. As one of our experts said: 'Grape ripeness can be a challenge for English wines, so it's encouraging to see two of them doing so well.'

To find out which supermarket Champagne beat the big brands and to discover Best Buy Prosecco and Cava for under £10, turn to p20. You can read full testing notes for all the wines on test at

From left to right: Marks and Spencer Marksman English Sparkling Brut NV Blanc de Blancs Vintage 2009 by Ridgeview (£26), Ridgeview Bloomsbury 2010 (£24.99) (Waitrose and online), Breaky Bottom Brut 2008 Sussex (£23.99) (selected branches of Waitrose, WaitroseDirect.com – minimum order of any 12 bottles), Plumpton Estate The Dean Brut NV (£20.99) (selected branches of Waitrose, WaitroseDirect.com – minimum order of any 12 bottles, and online), Sainsbury's Taste the Difference English Sparkling Brut 2007 by Denbies

Over the past twenty-five years, the viticultural landscape in Britain has changed dramatically

GB production of still and sparkling wine 2017-2024	2016	2017	2018	2019	2020	2021	2022	2023	2024
Still wines	31%	32%	31%	28%	36%	32%	32%	24%	31%
Sparkling wines	69%	68%	69%	72%	64%	68%	68%	76%	69%

Table 2

from one that was producing almost entirely still wines to one that today produces predominantly sparkling wines. WineGB surveys show that, on average, sparkling wine accounts for 69 per cent of production with still wines at 31 per cent. (See Table 2) One can see that in the very large harvest year of 2023, producers put more into long-term sparkling, presumably because they had plenty of still wines to sell and did not need any more. The reasons for the switch from still to sparkling are not difficult to see. The quality of the best sparkling wines made in Britain is, by any measure, considerably ahead of our still wines even though still wines are catching up. Our sparklers have elegance and finesse plus lasting power and have proved themselves in tasting after tasting to be the equal of Champagne and other fine sparkling wines. In addition, as I have already pointed out, they are much easier to explain to the wine-buying public and therefore easier to sell, especially through the on-trade. The early successes of brands such as Nyetimber, winners of major national and international awards with their first wines (from the 1992 and 1993 vintages) in 1997–8, and Ridgeview, whose first wine (a 1996) was released in 1999 and who won the Gore-Browne Trophy in 2000 (with their 1996 Bloomsbury) and in 2002 (with their 1998 Cavendish), plus the high prices they achieved, gave credence to the idea that sparkling wine was the way forward. The idea that 'because they can do it, so can I' – much in the way that the prominence of Lamberhurst Vineyards persuaded still winegrowers that they too could grow grapes, make wine and make money – persuaded many growers that sparkling wine was the product for the future. In addition, it was not very difficult to see that a bottle of Nyetimber or Ridgeview really was of an equivalent quality to a bottle of Moët, Lanson or Taittinger costing much the same, whereas a bottle of still wine from Three Choirs or Chapel Down was OK, but perhaps slightly (or even very) overpriced and therefore poor value for money. Couple this with the fact that bottle-fermented sparkling wine (aka Champagne) has the aura of prestige and quality about it and the average customer expects to pay more for it, and the appeal of sparkling wine to the producer from a financial standpoint is an obvious one.

Still wines fight back

When I wrote the first edition of this book in 2013–14, English and Welsh Sparkling Wine was definitely on a roll and it looked as if, given a decade or so, the switch from still to sparkling would accelerate to a point where still wine was perhaps only 10–15 per cent of production. (Mike Roberts, founder of Ridgeview, said he thought that sparkling wines would get to account for 90 per cent of all GB wines.) However, that appears not to be the case and while all the larger producers are still predominantly sparkling, many of them have now started to produce still wines, and many of the smaller new producers, who rely much more on local and farm-gate sales, have a mixed offering of both still and sparkling. We now know from the WineGB surveys that the still sparkling split is pretty consistent at 31:69 per cent.

The reasons are several and varied. Undoubtedly competition to sell English and Welsh sparkling wines through the trade, both on-trade and off-trade, has increased and there is only so much shelf space and so many hotel and restaurant wine lists that can find room for home-grown sparkling wines. Similarly, as competition has increased, the need to get customers to visit the vineyard, taste the wines, enjoy the ambience and – before they leave – put a couple of cases in the boot of their cars, has grown. Almost all of the larger vineyards have invested in visitor facilities in the last few years, ranging from simple tasting rooms and retail shops to very substantial offerings including catering facilities for groups, and even, in some cases, hotels and other on-site accommodation. This realisation that vineyard tourism has definite advantages and that retail sales direct to the consumer are much more profitable than trade sales, has meant that the wine offerings, and also the varietal spectrum, have changed. Varieties such as Bacchus, Ortega, Phoenix, Pinot blanc, Pinot gris, Sauvignon blanc, Siegerrebe, and Solaris for the production of still white wines, and Blauer Frühburgunder (Pinot noir Précoce), Regent, and Rondo, for the production of red and (more commonly) rosé wines have all increased their percentages of the overall planted area. Also rising up the charts are Cabaret noir and Divico, both red PIWIs, and Gamay which has always been grown in GB (the Castle Coch vineyards were planted with it in 1870) and it is being used now for both still and sparkling wines. In addition to the mainly 'still only' varieties mentioned in the previous paragraph, the percentage of still wines made from Chardonnay, Pinot noir and even Meunier, has risen considerably in the last fifteen years. Pioneers such as Chapel Down and Gusbourne, who to my palate made the best still wines from Chardonnay and Pinot noir, have been joined by others such as Danbury Ridge (see below), and while there are still a few 'sparkling only' producers, many of the major producers (apart from Nyetimber) now offer both styles.

Rise of the Crouch Valley winegrowing region

One major development in still wine production has been the rise of the Crouch Valley in Essex as a place to grow grapes. Why this has taken so long to happen is something of a mystery. I mostly grew up in the CM3 postcode, the postcode whose southern border is the river Crouch, and lived just north of Danbury between 1955 and 1970, attending schools and a college (only a technical college at the time, but now Anglia Ruskin University) in the area and know it pretty well. The CM3 postcode stretches from just north and to the east of Chelmsford down to the shores of the river Crouch between South Woodham Ferrers to the west and the outskirts of Burnham-on-Crouch to the East. When I told my parents (in 1973) that I was interested in planting a vineyard, my father insisted we visit New Hall Vineyard which was then almost the only vineyard in the area. I well remember the visit, meeting Bill Greenwood, who with his wife Shelia (and later his son Piers), had started the vineyard in 1969 with 850 Reichensteiner bought at auction by Bill (probably because they were going cheaply) and which they had subsequently expanded over the intervening decades to become one of the largest in the country. I do not recall tasting the wines on that first visit, although I do recall going down to the cellar under the house where their wines were stored at the time. I also remember that Bill wore both a belt and braces! Quite why that sticks in my mind I have no idea. I also remember the railway sleepers used as end posts, some of which are still in situ. Now they were a good investment! Over the intervening years, New Hall has won several major awards, including Winemaker of the Year in 2013, 2015 and 2016, but only five gold medals which for the size of the vineyard is a small number (Chapel Down for instance have won 46 in the same period). However, what New Hall also did was to establish themselves as one of the major suppliers of grapes to other vineyards and wine producers, including Lamberhurst, then by far the biggest in the country. New Hall had started growing Pinots in 1971, firstly Pinot gris (which they called Ruländer as the Germans still sometimes call it) and then in 1973, 1974 and 1975 planted Pinot noir in each year, a total of 2.33-ha (5.76-acres). Unbeknown to the Greenwoods, 50 Chardonnay vines had been mistakenly included among the Pinot gris, and when they delivered their usual loads of grapes to Lamberhurst in 1984, they included 1,300 kgs of Pinot noir and 150 kgs of Chardonnay, 'the very last grapes picked from the harvest' said Piers and asked Karl-Heinz Johner, Lamberhurst's German winemaker, to make them into sparkling wine. This wine was therefore almost certainly the first 'Champagne variety', bottle-fermented sparkling wine to be produced in Britain. Greenwood says the wine 'was released in December 1986 after ten months on the yeast' and is the wine which Chris Trembath, Piers's brother-in-law and now part-owner of New Hall Vineyards, remembers having at his wedding in 1987[9].

Since those early days, the number of vineyards in the Crouch Valley has greatly increased and the area planted with vines greatly expanded. In 2012 my vineyard database recorded that there were five vineyards in the CM3 postcode covering a total of 74-ha (183-acres). In 2015 this had risen to 8 vineyards with 117-ha (289-acres), and by 2024 to 16 vineyards covering 249-ha (615-acres) with an average size of 15.55-ha (38.42-acres) which is impressive. The other surprising element is that most of the land planted with finicky vines is clay, or more especially Essex clay. Thick and sticky one minute; hard as concrete the next. However, in the spirit of making the best of a bad job, the virtues of clay for vinegrowing have been fully explored and found to be 'perfect' for vines. Clay of course is not confined to Essex, and there are plenty of vineyards planted on Wealden Clay in Kent and East Sussex whose owners can now crow about their soils, rather than keeping quiet.

What has really inspired both producers and winemakers in and out of the region is the quality of the fruit, especially the Chardonnay, Pinot noir and Pinot gris. With its dry climate (some of the driest parts of the British Isles are in Essex) and its sheltered coastal sites, the sugar levels being achieved are some of the very highest in the country and 14½ per cent natural alcohol is not unheard of for Chardonnay. This level of quality has attracted a lot of attention from grape buyers and persuaded some of the larger producers in other parts of the country (Balfour, Chapel Down, MDCV) that this is the place to plant. The latest (and very high-profile wine producer) to plant here is Jackson Family Wines who paid what appears to be an extraordinary amount of money to obtain land in the area as their first foray into the world of English wine. And planting is not only confined to the CM3 postcode with land being sought and planted south of the River Crouch in the area between the Crouch and its tributary river the Roach going south towards Southend. The total number of Essex vineyards in 2015 was 25 covering 144-ha (356-acres); by 2024 this had almost trebled to 66 vineyards covering 417-ha (1,030-acres).

The best known of all the CM3 producers, apart from New Hall which has the pedigree of age which cannot be replicated, is Danbury Ridge Wine Estate, planted in 2014, who in less than a decade, have managed to achieve an enviable position for the quality of their still wines made from Chardonnay and Pinot noir. Other large growers are: Crouch Valley (the name of an individual vineyard and wine producer), Clayhill, Crow's Lane Estate, Great Whitmans, Martin's Lane, Missing Gate, New Hall, Riverview, and Trewince Estates.

This emergence of a region whose principal output is seemingly mostly used for the production of still wines comes as something of a welcome relief from all that land planted for

9 See Annex I for the full history of sparkling wine production in GB.

sparkling wine and nothing else. If the quality of the best presently coming out of wineries using Essex Chardonnay can be maintained and maybe even improved, then this will be a very welcome addition to the spectrum of wines being produced. I have long maintained that if growers can consistently make still wines that sell at the same sort of prices as good Chablis and white Burgundy, then this will do a lot for the sustainability, that is financial sustainability, of the whole industry. Selling at between £35 and £60 a bottle, with relatively short ageing required (compared to the years that top quality sparkling takes), and with no expensive riddling and disgorging, these wines are undoubtedly profitable. Let's pray that global warming continues to allow us to produce wines like this and even better.

Production and sales of all GB wines -in millions of 75 cl bottles

Years	**Ha in production**	**Production**	**Sales**
2005-2009	779	2.19	N/A
2010-2014	1,296	3.78	N/A
2015-2017	1,648	4.83	1.4 (average)
2018-2121	2,414	10.09	6 (average)
2022	3,103	12.18	8
2023	3,661	21.60	8.8
2024	3,763	10.67	9.1
2027	4,553	17.60	13
2029	5,509	21.30	16

2027-29 figures based on estimated increase in plantings of 10% per annum & average production (2015-2024) of 29 hl-ha. 2027 & 2029 sales estimates @ 10% pa increase

Source: Wine Team at the Food Standards Agency and WineGB Industry Surveys

Table 3

Oversupply of English and Welsh sparkling wine?

Another factor that ought to be taken into account when considering planting a vineyard is the question of oversupply. Since the dark days of the last century, when the area under vine decreased by 30 per cent as growers grubbed up their old, mainly Germanic varieties, the area under vine has risen from a low point of 761-ha (1,880-acres) in 2004 to today's estimated total of at least 5,000-ha (12,355-acres) and probably more. And, of course, with the increase in vineyard area has come the inevitable increase in production. As can be seen from Tables 3 and 4, the average annual production of all wines has risen from 2.19 million 75 cl bottles per year in the five years between 2005 and 2009, to a frightening average of almost 15 million bottles per year between 2022 and 2024. However, these figures are based upon 'area in production' and not total planted area. The latest area in production figure available is for 2024, and is 3,763-ha, with a further 982-ha of vineyards which are three years or younger and 'not in production', a total of 4,745-ha. This means that by 2027, taking the actual current total planted figure and using the latest 10-year average yield of 29 hl-ha, the yield could be 13 mn bottles by 2027, and 16 mn by 2029. Of course, if the yields were higher than average (as they were in 2018 and 2023 – average 45 hl- ha) then the 2027 figure could be nearer 19 mn bottles, and the 2029 figure 21.50 mn bottles.

Production of still and sparkling wine 2016-2024 - Mn x 75 cl bottles

	2016	**2017**	**2018**	**2019**	**2020**	**2021**	**2022**	**2023**	**2024**
Still wines	1.28	1.63	4.03	3.23	2.71	2.75	3.75	6.65	3.28
Sparkling wines	2.87	3.65	9.08	7.25	6.10	6.20	8.43	14.95	7.38
Total	**4.15**	**5.28**	**13.11**	**10.48**	**8.81**	**8.95**	**12.18**	**21.60**	**10.67**

Source: Wines of Great Britain annual producer surveys

Table 4

Sales and distribution of GB wines 2017-2024 in millions of 75 cl bottles								
Sales	**2017**	**2018**	**2019**	**2020**	**2021**	**2022**	**2023**	**2024**
Total sales	1.4	2.6	5.5	7.1	9.3	8.0	8.8	9.1
Still wine	0.46	0.73	1.82	2.56	3.44	2.48	2.64	2.91
Sparkling wine	0.94	1.87	3.69	4.54	5.86	5.52	6.16	6.19

Source: WineGB Industry Surveys. 2017 and 2018 figures are industry estimates

Table 5

Sales by sector of GB wines 2017-2024									
Sales	**2017**	**2018**	**2019**	**2020**	**2021**	**2022**	**2023**	**2024**	**Averages 2019-24**
Cellar door	36%	32%	29%	28%	38%	19%	17%	19%	27%
Trade sales	55%	54%	54%	46%	39%	63%	64%	60%	54%
Exports	4%	8%	10%	4%	4%	7%	8%	9%	7%
Online	5%	6%	7%	22%	19%	11%	11%	12%	12%
	100%	**100%**	**100%**	**100%**	**100%**	**100%**	**100%**	**100%**	**100%**

Source: WineGB Industry Surveys. 2017 and 2018 figures are industry estimates

Table 6

Sales of English and Welsh wines

Nobody really knows for sure how many bottles of English and Welsh wines, both still and sparkling, are sold each year. The best data available is taken from the WineGB's 'Industry Reports for the years 2019-23. These reports are based on a voluntary survey sent out to both WineGB members and non-members with data submitted anonymously. Sales, as can be seen in Table 5, are increasing, but nowhere nearly as fast as production is. As the table shows, sales rose from 5.5 mn 75cl bottles in 2019, peaked at 9.3 mn during the Covid lockdown period, fell back in 2022, but are rising again. Sales in 2023 and 2024 rose very slowly, reflecting the cost-of-living crisis which has affected sales of all wines, especially high-end sparkling wines. Current estimates of sales for 2025 are around 10 mn bottles which is still only around 65 per cent of current production meaning that stocks of wines, especially sparkling wine, are still rising.

Sales by sector

Table 6 shows the distribution of sales of English & Welsh wines by sector. As was stated previously, these figures are based upon voluntary surveys with the data given anonymously. As can be seen, cellar door sales by volume have fallen, as trade sales have risen, no doubt owing to larger producers having reached some sort of limit on their capacity to sell more wine from the vineyard gate as more individual producers open up to visitors and direct sales. Exports rose until 2019, then fell during Covid, but are back on the rise again. Online sales also followed much the same trajectory as cellar door sales, rising until Covid, then falling back as trade sales rose. As the number of producers who supplied data (146) was quite small, there are bound to be some anomalies, but it is the best we have at the moment. Given that many of the producers who responded were the larger ones, more given to trade sales and exports than smaller ones are, these percentages are almost certainly overstated for the whole industry. It is my guess that what the WineGB survey revealed was that of those producers who export, they export around 7-8 per cent of their total sales. On an industry-wide basis, therefore, exports account for maybe only 2–3 per cent of the total. Good data from the HMRC Excise Duty returns used to be available, but since the new duty rates and arrangements have changed, these are no longer available. A pity, although in any event it did not include exports and other duty-free sales (airports, ferries, cruise lines, private exports to consulates, and other duty-free sales). HMRC could also not account for domestic sales into GB bonded warehouses. These sales went in as being of 'UK origin' but appeared to come out lumped together with all other sales. It is a great pity that HMRC cannot help in this respect, especially since GB sales raise almost £40 mn for the government from Excise Duty and VAT alone.

Top: Figure 1
Bottom: Figure 2 Source CIVC

Stocks of English & Welsh wine

No official data exists for stock levels of GB wines, something which I feel WineGB ought to be collecting, as it is a valuable figure and helpful for producers to have when estimating future sales and competition. This lack of reliable data means that estimating stock levels is an art, rather than a science, but it is useful to make a stab at the figures. Taking the available data on production and sales, I have calculated that, at March 2025, stock levels of sparkling and still wine in Britain are probably around 60 million bottles with sparkling alone at around 40-50 mn bottles. If we accept that sales will rise to around 10 mn bottles a year of which 67 per cent is sparkling, then this is almost 7½ years' sales worth of sparkling wine. While this might sound a lot, of course, as has been pointed out earlier, if you are in the sparkling wine business, then by definition you need stocks

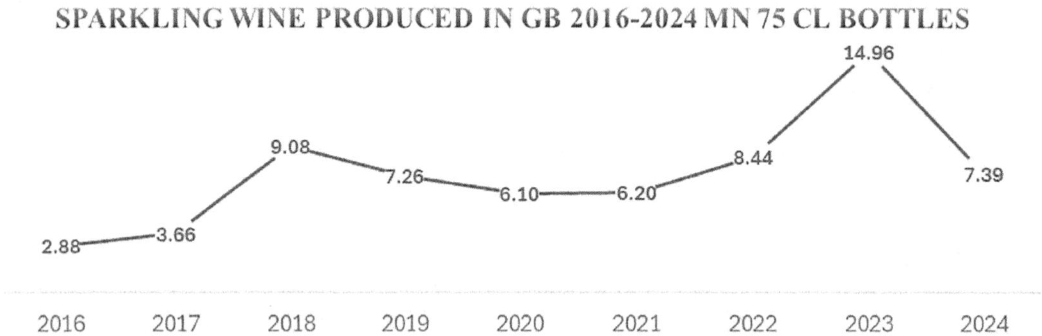

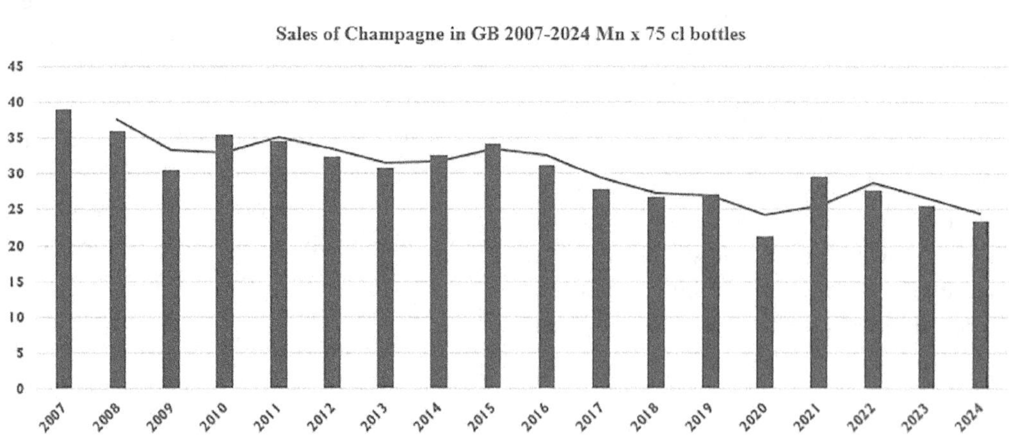

of wine and, cash-flow issues aside, stocks of four, five and even six years' worth of sales are nothing unusual. In Champagne, the average stock levels of all wines, non-vintage and vintage, currently amount to nearly five years' worth of sales. This includes all the minimum-aged Champagnes with attractive sounding (but mainly made-up) names that are the staple offering of supermarkets around the world which spend a bare 15 months *sur latte*. However, if you look at *Cuvée de Prestige* and *Millésimés* Champagnes (which together account for around 6 per cent of total production by volume), then the stock levels are much higher. Current releases of vintage Champagnes range from the early 2000s to 2016, with nothing much younger than that, in other words at least seven to eight years' worth of stock and for many brands considerably more. In this context, therefore, stocks of (mainly vintage) English and Welsh sparkling wines of six to seven times annual sales do not seem that abnormal.

Where will it all sell?

Given the production and sales figures already shown, the question of 'where will it all sell?' is given some perspective. Sales in all markets, direct to consumer, trade and export of GB sparkling wine in 2023 were 6.16 mn bottles and in 2024 were 6.19 mn bottles, so hardly increasing. Judging by the current financial situation and the state of the hospitality industry, the feeling among producers is that they are unlikely to rise much in 2025. Some producers think that exports might rise, but as this only represents around 9 per cent of sales (which, as I have said before, is probably and unrealistically high number) even doubling this would not dent stocks too much. So, let's say that sales in 2025 reach 10 mn of which 65 per cent is sparkling. 6.5 mn bottles is still less than half of the 15 mn bottles a year that have been produced between 2022 and 2024 and we know that production will rise over the next two to three years as the 1,000-ha of vineyards already planted but not yet producing start croping and their wines from early vintages mature and come into the market.

The good news is that English and Welsh wines of both types continue to sell, and the sales figures are at least showing modest increases. More vineyards are coming on-stream and finding new markets to sell their wines to. When you consider the overall size of the wine market in GB, which even with Covid, Brexit, inflation, duty changes, the rise of the 'no-and-low' drinks sector, and the war against alcohol being carried out (probably quite sensibly) by the medical and anti-alcohol world, it is amazing that sales of home-grown wines are increasing at all! However, as more consumers get used to the idea that English and Welsh wines offer a quality alternative to wines from abroad, and as more consumers see our wines on pub, restaurant, club and hotel wine lists and they feel reassured about the quality, then they will buy them. When you look at how Champagne sales have fallen since 2007 (Figure 2), the year before the banking crash of 2008, it is plain to see where some of the lost sales have gone. From a high of 39 mn bottles sold in GB in 2007, sales of Champagne have fallen to an estimated 24 mn bottles in 2024, their lowest figure for decades. Of course it is not all down to GB wines. Prosecco, *Cremant* and even Cava, plus other sparkling wines of all origins and types have taken a nibble out of Champagne sales, especially as the price of Champagne has risen.

Still or sparkling? A summary

The answer to this question I believe is rooted in several factors. When visiting ('interviewing', I was tempted to write) new entrants to the business of establishing a vineyard, growing grapes and selling the produce, I always ask the three W questions: What, where, and who? What are you proposing to sell? Where are you going to sell it? Who is going to sell it? These three questions are really at the heart of a successful vineyard enterprise. If the answers to these three basic questions is: the finest quality English wine ever made in this country; I will employ a great salesperson; and 'high end hotels and restaurants', then it is time to rethink the whole enterprise. As much as someone might want to run their ship like this, it doesn't often work out that way. Other questions such as: are you going to be selling all of your grapes? Some of them? None of them? Will you be selling wine? If so, what type and where? Direct to the public? To wholesalers and retailers? Hotels, pubs and restaurants? You would be surprised how many people have given this aspect of their future enterprise very little thought. However, from the answers to these questions stem a lot of the answers to: 'what do I plant?'

There is no doubt that where, as an industry, we once all thought that sparkling wines would come to dominate, there are very positive signs that still wines, with their infinite variety of styles, brought about by using different grape varieties and different winemaking techniques, can play an important part in a vineyard's wine offering. Apart from any other consideration, if only for cash-flow reasons, still wines can play an important part in creating a viable business. Still wines can start to return you an income within three years of planting; with sparkling wines it might be five or six. At the volume end of the GB sparkling wine business, currently occupied by around twenty sizeable producers (but surely to be joined by more to come), my guess is that, across their range, most of them will have quite marked price differences, with prestige cuvées, middle-rank brands and (when necessary) 'on promotion' offerings all under the umbrella of the same label and/or brand. Nyetimber – as an example – currently have their '1086' brand at £130–£150 a bottle; various vintage wines around the £35–£45 mark; and their 'multi-vintage' widely available at £28–£30. I do not suppose the cheaper offering stops people buying their more expensive wines, any more than Moët & Chandon's offerings do across different price bands. In addition, the last 5-10 years have seen the introduction of different types of sparkling wine. Charmat method and carbonated wines, in bottles and in cans, whether you like them or not, are here to stay, if only because of their lower price points. They are also popular with some producers as the production costs are lower, stock levels are lower, and cash-flow is improved. We have also seen such things as orange wines, natural wines, pét-nats, *Col Fondo* wines, wines in cans and novel bottle shapes, types and sizes as well as some no-and-low offerings based on GB wines. And why not? If it keeps vineyards in business, so be it. We have also seen the rise of independent winemakers setting up wineries in both urban and rural locations, on farms, on trading estates, under railway arches, buying grapes from growers they can work with, and on the harvest-time spot market (which is growing in size) and making good quality and inventive wines. This is yet another outlet for the 33 per cent of grapes produced that get sold each year[10].

When it comes down to the wire, deciding what varieties to plant, and whether to produce still or sparkling wine, is mainly about your expectations and your budget. If you are proposing to have the all-singing, all-dancing, full-on visitor experience vineyard, then I would suggest planting a spread of varieties to cover both still and sparkling wines, and to create as broad a wine offering as possible in order to satisfy as many palates and as many budgets as you can. If your aspirations – and wallet – run to a bigger operation where, by definition, your sales cannot (or are unlikely to in the medium- to long-term) be mainly over the farm-gate, then establishing a sparkling wine brand, which, unless you are very lucky, will take a minimum of ten years, and probably a deal longer, may well be the way forward. As ever, financial considerations are as important as viticultural ones.

10 33 per cent is an average across the last three years, 2023-25, which showed that between 26 and 38 per cent of grapes were traded. These could be both grapes under contract and open-market sales.

Chapter 2

Recommended vine varieties for Great Britain

Note 1: The vinifera **varieties recommended in this chapter are for those suitable for COOL climates such as that found in the southern half of Great Britain. This means climates that meet the following parameters:**

- An average maximum temperature of +/- 15°C
- GDDs of between 900 and 1,100[11]
- Between 1 and 15 days in the year when the temperature reaches 30°C or more
- Climates where temperatures seldom fall below -10°C

Note 2: After the sections on major varieties, there is a section for **Vine varieties for less favourable sites**

Note 3: Full details of PIWIs and hybrids can be found in Chapter 9

Note 4: The varieties in this section are listed in order of quality, although when you get into the less-widely grown varieties much depends on the quality of the site and the capabilities of both the viticulturalist and the winemaker.

Still white wines

If you have the right site, then **Chardonnay** for still wines is king of the castle. It has been discussed before in this book (see under Crouch Valley section in Chapter 1) and on suitable sites and in ripe vintages, it can be very good and can sell for high prices. However, it would be a brave grower, especially if they were new to the business and unsure of the quality of their site, who based their business plan on growing Chardonnay solely for the production of still

11 Growing Degree Days between 1 April and 31 October using 10°C as a base. No cut-off at 30°C

GB - Yields of major white vine varieties 2016-2024

Tonnes per active Ha	2016	2017	2018	2019	2020	2021	2022	2023	2024	Average 2016-24	Average tonnes-acre
Bacchus	3.07	4.21	5.48	7.14	3.48	5.91	6.70	9.92	6.20	5.79	2.34
Chardonnay	5.58	4.42	8.67	6.61	6.33	4.75	5.54	9.83	4.30	6.23	2.52
Alzey 7672	6.05	5.64	6.70	6.44	5.22	5.05	8.28	7.17	3.50	6.01	2.43
Ortega	No data	No data	No data	6.53	3.38	3.75	6.30	8.41	3.30	5.28	2.14
Pinot blanc	No data	4.86	7.85	No data	No data	No data	8.88	7.60	3.10	6.46	2.61
Pinot gris	No data	No data	No data	2.21	4.02	4.82	5.26	10.46	3.50	5.04	2.04
Reichensteiner	8.36	4.37	16.60	7.50	2.34	N/A	8.20	8.70	3.20	7.41	3.00
Seyval blanc	6.40	11.24	9.38	3.36	4.61	7.39	6.73	9.13	2.50	6.75	2.73
Solaris	No data	No data	No data	2.66	1.99	4.39	4.90	6.45	1.70	3.68	1.49
Averages	**5.89**	**5.79**	**9.11**	**5.31**	**3.92**	**5.15**	**6.75**	**8.63**	**3.48**	**5.85**	**2.37**

Source: Data collected by S. P. Skelton from growers (2016-17) and data collected for the WineGB annual yield report (2018-24)

Table 7

wine. But Chardonnay is an adaptable variety and if you are normally producing sparkling, then in good vintages you can select your best parcels and attempt a still wine. Of course, good Chardonnay often requires barrel fermentation and/or barrel ageing and, given its typically high acidity, it also needs time to be at its best. However, if you can produce a Chablis-style unoaked or lightly oaked wine and sell it within 24 months of harvest at a decent price, then you will have a very profitable product.

Next, in terms of pure wine quality I would recommend **Pinot gris.** Again, it needs a good site to get really ripe grapes and to maintain reasonable yields. The quality of the wines can be high, even very high. Pinot gris is a name that people recognise and with great examples coming out of many other countries, Germany, the Alto Adige and Veneto in Italy and of course Alsace, it is variety that people know and trust. In less ripe years of course it can be added to sparkling cuvées. However, yields even on good sites, can be variable.

Pinot blanc is also a variety to consider. Again you need a good site to ripen it fully and get the acids down but it makes very good wines in a neutral Chardonnay style. It also blends well with other lower-acid aromatic varieties. It takes oak well and of course can be used for sparkling.

For most still white wine producers, **Bacchus** is a 'must have' variety, especially if you have a site that really suits it, you know how to grow it and it ripens well. The trade and the public recognise it now as a 'British' variety and given its success in tastings and with a great marketing name, its future is secure. On good sites in the right part of the country, yields can be high without a reduction in quality.

Seyval blanc has made a modest comeback in the last decade – it is fifth in planting area which at 123-ha is the highest it has ever been since records began – and its above-average yields, plus its natural disease resistance make it a reliable variety. It can produce good stand-alone wines – wines made with it have won the Gore-Browne Trophy three times – and it has also made some great sparkling wines. It also makes a very good blending partner with other varieties, neutral or aromatic, and for those either making Charmat or carbonated wines, or selling their grapes to producers that do, I would definitely recommend it. For those in the more climatically challenged parts of GB, it is a valuable variety.

Reichensteiner is also a variety worth considering, especially if you are not in prime vineyard country, and/or your economic model requires good yields (the highest of any variety – see Table 7) with low inputs. It has also been used successfully in sparkling wines and if picked with sufficient acidity, can be very good. It is also excellent for blending, especially with higher acid, aromatic varieties.

Next in terms of quality I would rate **Ortega.** Named after a Spanish philosopher, it is an aromatic, low-acid variety that is tricky to grow, has an early bud-burst (and is thus susceptible to spring frosts), ripens early so gets attention from wasps and is prone to *botrytis*. But its wine quality can be excellent, and the fruit flavours are like no other variety and is also good for blending. Plantings have been expanding slowly but surely over the years and once you work out how to look after it, it is valuable variety to have on the tasting table.

Solaris finds favour in the more marginal regions and the area planted has expanded in percentage terms more than any other still wine variety. The total planted now stands at 117-ha, the sixth most widely planted variety. Having tasted many different examples from all over Europe, especially from Baltic Germany, Poland and the Scandinavian countries, I am convinced that it can produce very good wines, both still and sparkling. However, in GB it is mainly planted in the west of the country, Wales and the northern regions where vines are grown and where qualities are generally lower when compared to the east and south-east. Of the 117-ha grown in GB, there are around 40-ha in Kent and West Sussex, and the remaining 77-ha are elsewhere. This does mean that the variety has had something of a chequered career, although it has won several good prizes in competitions, both at home and abroad. If having a variety that ripens early, gets high sugars and has some natural resistance to diseases are important, then this is the variety for you.

Of all the other varieties grown in GB (a total of 92 different varieties) there are very few that can be recommended for still white wine without reservation. **Schönburger** and **Siegerrebe** can both produce very good wines in a certain style, but they are difficult varieties to sell, especially through the trade and I would only recommend planting them if the vineyard proposes to sell these wines mainly from the farm-gate. If you had a very good site, you might test the water with **Albariño** (called Alvarinho in Portugal), but it takes quite a bit of ripening, so a good site, lots of shelter and patience are needed. It is however a variety that people recognise and is easier to sell than a variety with an ümlaut in the name (although it does have a tilde). **Phoenix** has its fans, but it has not increased in planted area much in the last 20 years and I cannot see it really taking off now. Better to grow Bacchus if you like that style of wine. There are a few producers who have made white wines from **Pinot noir** and while there were very few examples that I thought worked, experience has paid off, and there are a few that are now worth drinking. Maybe with time 'White Pinot' will become a common sight on the wine shelves. And I guess **Meunier** might also be bracketed with its black cousin in this respect, although white wines from this variety are very rare.

Of the minor varieties, **Optima** and **Orion** might be interesting on a good site, and their wine quality is good. I have a soft spot for **Auxerrois, Gutenborner** and **Regner,** varieties that have done well in the past (I won a gold medal with a Gutenborner in 1983) and from which very good wines can be made. As for other varieties, there are precious few that will make money. There are some growers who believe that **Sauvignon blanc** can be successful in Britain and maybe, in years to come, they could be right. However, at the moment, except on very good sites and in exceptional vintages I disagree, and would not recommend it, at least, not as a mainstream variety. Woodchester Valley Vineyard, near Stroud in Gloucestershire, won an IWSC Gold and a Gold at the Drinks Business 'Global Masters' competition for their 2018 Sauvignon blanc, which shows that it can be successful, but which rather goes to confirm my point about it needing an exceptional vintage for it to ripen. Having said all of the above, I was intrigued to see that in 2025 that the sole vineyard parcel planted with the Rhône variety **Marsanne** actually produced a crop of ripe grapes! However, it will not be featuring (yet) on my list of recommended varieties.

Still white PIWIs and hybrids

Of all the many PIWIs and hybrids being grown in GB, the only white varieties that I can wholeheartedly recommend are **Solaris** (already mentioned above), **Calardis blanc, Calardis Soleil, Muscaris, Ravel blanc, Rinót, Sauvignac, Seyval blanc** and **Souvignier gris.** The others are works-in-progress (especially the reds) and until someone proves otherwise, be cautious. See Chapters 7 and 8 for much more on PIWIs.

Varieties NOT recommended

Older varieties that were once more widespread such as **Faberrebe, Huxelrebe, Kerner, Kernling, Scheurebe,** and **Würzer**, while they linger on in some of the older vineyards, have all had their day and nobody is planting them anymore. In no circumstances would I recommend **Riesling.** We just do not have the climate for it (yet) and despite some large plantings in the past (mostly now grubbed) it has never performed well. I have already discussed **Sauvignon blanc** and cannot recommend it, but I will probably be proved wrong in 20 years' time.

Note: For more details on each recommended variety, see individual entries in the next chapter.

Still red wines (including red PIWIs)

For still red wines – if REAL quality is the aim – the choice is very limited. The best two are **Pinot noir** and **Blauer Frühburgunder** (Pinot noir Précoce), both of which, given a good site, the right crop load, perfect canopy management, good disease control and, of course, good winemaking, can make world class red wines. I cannot pretend that they always match the finest wines from Burgundy, Martinborough, California or other Pinot hot-spots, but the best from Britain have an excellent purity of fruit and a refreshing acidity that is sometimes lacking in wines from warmer climates. Whether they are value for money is another matter. The best sell for anywhere between £20 and £45 a bottle, which is pricey for most customers but as there are (almost) no good cheap Pinot noirs from anywhere, why should ours be any different? In the national 'Wine of the Year Competition' wines made from Pinot noir have won the 'best red' trophy a total of 23 times out of the 33 times it has been awarded, with Beenleigh's (tunnel grown) Cabernet Sauvignon and Merlot blend winning it 4 times, and wines made from Rondo (4 times) and Dornfelder, Pinot noir, Regent and Rondo blends the other years. Since the competition started in 1974, wines

GB - Yields of major red vine varieties 2016-2024

Tonnes per active Ha	2016	2017	2018	2019	2020	2021	2022	2023	2024	Average 2016-24	Average tonnes-acre
Früburgunder, Blauer	1.81	3.50	6.62	4.32	3.55	5.34	5.59	6.97	1.80	4.39	1.78
Meunier	5.88	5.24	9.53	6.23	5.36	4.66	6.56	10.49	3.50	6.38	2.58
Pinot noir	3.00	4.31	7.93	6.18	5.25	5.70	6.01	9.20	3.90	5.72	2.31
Regent	No data	5.70	11.09	4.12	3.10	6.25	8.09	9.94	2.50	6.35	2.57
Rondo	No data	9.24	7.30	2.66	5.17	6.09	7.01	9.18	2.60	6.16	2.49
Averages	**3.56**	**5.60**	**8.49**	**4.70**	**4.49**	**5.61**	**6.65**	**9.16**	**2.86**	**5.80**	**2.35**

Source: Data collected by S. P. Skelton from growers (2016-17) and data collected for the WineGB annual yield report (2018-24)

Table 8

made with Pinot noir have won either gold or silver medals a total of 56 times, a total far, far higher than any other red variety. A wine made from Frühburgunder has never won the 'best red' trophy, but wines made from it have been awarded a total of 14 gold and silver medals.

Red vine varieties 2009 & 2024 in hectares

Type	Variety	2009	2024
vinifera	Pinot noir	217.97	1,369.77
vinifera	Meunier	48.70	410.78
vinifera	Frühburgunder, Blauer	17.31	95.00
PIWI	Rondo (GM 6494/5)	43.00	64.23
PIWI	Regent	19.30	32.42
vinifera	Dornfelder	18.22	15.85
PIWI	Cabaret noir	0.00	14.75
PIWI	Divico	0.00	14.03
vinifera	Gamay noir	1.23	7.69
PIWI	Cabernet Cortis	0.00	6.57
vinifera	Acolon	4.59	5.75
PIWI	Pinotin	0.00	3.58
vinifera	Merlot	1.13	3.06
vinifera	Zweigeltrebe	1.34	1.21
vinifera	Cabernet Franc	0.02	1.01
Totals		**372.81**	**2,045.70**

Source: Wine Team at the Food Standards Agency

Note: Area stated for Früburgubnder is an estimate

Table 9

Having said the above about how good wines made from Pinot noir and Frühburgunder can be, there are sadly far too many red wines (and some of them not terribly red) made from them in Britain that are not great quality, and they would be far better off as rosé wines. In my experience, many red wine drinkers are looking for heavy, full-bodied ripe tannin-rich reds, wines that with the current varieties we grow, and the current climate, we can rarely produce. Unless you know your site well and are sure that you can ripen Pinot noir and Frühburgunder to the level required to make a really decent red, I would guard against basing my whole business plan on producing still red wine from these varieties every year. By all means try and in good vintages you might succeed but treat it as a bonus when you do. If in doubt, make a rosé.

As for other red varieties, I believe that **Meunier** might surprise everyone and turn into Britain's USP for still red (and rosé) wines plus of course for sparkling wines of all types. It also has the advantage (for the grower) that it yields well – highest yields of the top four varieties, white or red – and a name that wine buyers will have heard before.

The next three most widely planted reds, **Rondo, Regent,** and **Dornfelder** (in order of planted area) are all what today are considered old-school varieties. If you have them in your vineyard you are probably not going to grub them, but they are not being planted widely any more. **Rondo,** which ripens early, has large crops, low acidity and excellent colour, is being planted in some of the more challenging regions of Britain, but is now less common in the southern half of the country. **Regent** is one of my recommended PIWIs and full details can be found in the chapter on PIWI varieties. **Dornfelder** is now being grubbed and the area under vine is falling, although only slowly. If you have a vineyard open to the public and intend to major on farm-gate sales, then a red wine made from a blend of Dornfelder, Regent and Rondo will probably sell and will certainly have better colour than any Pinot-based wine, but they are not varieties I would recommend for good sites or for maximising your reputation or your bottom line.

The next two most widely planted varieties to consider, **Divico** and **Cabaret noir** (and not Cabernet noir which is a name the EU rejected) are both very new to the GB wine scene. They appeared in the 2022 list of varieties being grown in GB and have expanded quite quickly with each now having around 14-ha planted. To date, very few single varietal wines have been made from these, and the jury is out as to whether they are the answer to a maiden's prayer or will go the same way as Rondo and Regent, both touted as 'game changers' when they were released. They are both PIWIs, so will appeal to those seeking varieties needing lower pesticide inputs. Their yields also ought to be good in the right hands.

Of the other varieties that are planted on 1.0-ha or more, **Gamay noir** has its followers and has been planted in the past (the oldest block still extant was planted in 1984) and there are a few new growers, but I really cannot see the benefits of planting it. While it might make a fair sparkling wine, as a red or rosé I do not believe it shows any advantages over Pinot noir. Of the others, **Cabernet Cortis** and **Pinotin,** both fairly recent PIWIs, might be worth experimenting with. **Merlot** I would not give much space to – why would you? **Zweigeltrebe** is another variety I would not recommend as apart from an impossible name, it needs more heat than we usually have in GB to make even a half-decent rosé, let alone a red. And as for **Cabernet Franc** – again, why would you?

There are an additional seven other red PIWIs being grown in GB vineyards in quantities of less than 1-ha, some of which might be worth trialling. Please take a look at the individual variety entries in the chapter on PIWIs for more details.

Still rosé wines

Still rosé wines can be commercially very successful in Britain, and I wish that some of the growers currently striving to make palatable red wines would instead turn their **Pinot noir** and **Blauer Frühburgunder** (Pinot noir Précoce) into attractive, fruity rosés. The Pinots are to my mind the best varieties for still rosés and have the advantage that in poorer years they can go into sparkling wines and in really ripe years an attempt at a decent red can be made. They also have the advantage of a recognisable name, a fashionable light-red to grey-tinged *oeil de perdrix* (partridge eye) colour and on the palate live up to the consumer's expectations (which is generally not very high or demanding for rosés). Rosés made from other red varieties – **Dornfelder** or **Regent**, for instance – or from blends of white and red varieties can also be successful, but none to my mind have quite the class to match Pinot noir. Again, as with red wines, varietal naming of rosés is not so important, and a multi-variety rosé can work well. I would avoid the bulk of a rosé being made from aromatic varieties – Bacchus, Schönburger, Ortega, Siegerrebe – and use a more neutral variety such as Seyval blanc, Reichensteiner, maybe even Alzey 7672, or Pinot blanc if your site will ripen it, and colour it with one (or more) of the better reds.

Some of the PIWIs would work well as rosés, so please see the entries on **Cabernet Cortis, Pinotin, Prior** and **Regent** in the chapter on PIWIs.

White traditional-method sparkling wines

For traditional bottle-fermented white sparkling wines, the Champagne trio of **Chardonnay, Pinot noir** and **Meunier,** either singly or in blends, have proved themselves more than capable of producing world-class wines on many sites in Britain. *Blanc de Blancs* made from 100 per cent Chardonnay have been particularly successful and are amongst the very best wines that we can produce. However, they do not suit all sites and I would caution against planting these higher acid varieties (especially Chardonnay) on challenging, i.e. marginal, sites. In difficult and late years, acid levels can remain stubbornly high, although with time, even cygnet-wines with high acid levels can age into graceful swans. On marginal sites, of course, yield levels will be adversely affected which will have an impact upon incomes. Having praised the Champagne varieties, I would not discount using other varieties such as **Auxerrois and Pinot blanc** especially if used in blends.

I will reserve special praise for one of GB's longest serving varieties, **Seyval blanc**, 75 years and counting since it was first planted at Hambledon in 1952. As a single (or dominant) variety, **Seyval blanc** certainly has its place in sparkling wines made in Britain, although it is usually (but not always) going to lag slightly behind the Champagne-variety based wines in terms of quality and retail price (although some producers – Breaky Bottom, Camel Valley, Denbies – are very successful with it and their wines fetch good prices). However, the ease with which Seyval blanc can be grown and the level of yields it typically produces may well be more than adequate compensation for these shortcomings.

Although there have been a few acceptable sparkling wines made from varieties such as **Reichensteiner** and **Pinot gris** usually in blends but occasionally as solo varietal wines, I do not believe anyone planting today should choose these varieties solely for this style of wine. As for other varieties, I would personally avoid any of the aromatic varieties and, worldwide, these seldom make classic sparkling wines (Riesling Sekt for instance). However, there have been a few sparkling wines made from varieties such as **Bacchus** and **Solaris**, but I do not think they have the same gravitas (or sell at the same prices) as wines made from more 'noble' varieties. The advent of Charmat and carbonated wines in bottles and cans, where the wine is sold more on price and style than pure quality, has opened up the possibility of making sparkling wines out of more-or-less any of the varieties being grown in volume in GB today. For these wines, my advice would be to select a variety with a high yield potential. **Seyval blanc and Reichensteiner,** plus maybe **Solaris** spring to mind. See section below on non-traditional-method sparkling wines for more discussion on suitable varieties.

Rosé and red traditional-method sparkling wines

For rosé sparkling wines almost any combination of Champagne varieties works well, as do single variety wines made from either **Pinot noir** or **Meunier.** All can be successful. For many years Meunier was discriminated against (my theory is that the French persuaded Nyetimber that the variety could not make good wine, so no one coming after them bothered with it) but in recent years it has produced some excellent wines, both rosé and red wines, still and sparkling. There is more on Meunier in the individual variety description in the chapter on single varieties. For the red element of rosé wines, I prefer wines where the colour has come from red-fermented Pinot noir or Meunier. I can understand why some growers use a small addition of wine made from Blauer Frühburgunder (Pinot noir Précoce), or one of the other red varieties such as **Dornfelder, Regent** and **Rondo,** although I do not think they are as good and would not advise them for a new vineyard. Other neutral sparkling wine varieties – **Seyval blanc, Pinot blanc, Auxerrois** – can be used for blending into rosé wines, but they are rarely as good. Rosé sparkling wines often have their colour tweaked at the time of disgorging and *dosage* and in my experience, red Pinot noir works best, although I have known wine from other high-colour varieties used successfully.

As for sparkling red wines, to date there have been very few that could be called 'red' – a few dark rosés perhaps. However, I personally quite enjoy semi-sweet sparkling reds – some of the Australian Syrah-based ones can be very good and I have also had a Nebbiolo-based red sparkling wine from Asti which was excellent – so I see no reason why **Dornfelder, Rondo** and **Regent** could not be used, or maybe one of the newer PIWIs such as **Cabernet Cortis, Cabaret noir, Divico, Pinotin,** or **Prior.** I would prefer to see these wines in the *demi-sec* or even *doux* category as British acids and tannins would normally be too high for bone-dry red sparkling wine.

Novel styles of still & sparkling wines

Carbonated, tank-method (Charmat), Pét-nat, Col-Fondo, natural, orange and flavoured wines. Plus vermouth and spirits.

When I wrote the first edition of this book in 2012–13, non-traditional-method sparkling wines barely got a mention. Sure, there had been carbonated sparkling wines produced in the past (the carbonated 1983 Barton Manor Sparkling Rosé won a gold medal in the 1984 English Wine of the Year competition) but they were few and far between and barely got noticed outside of the English wine fraternity. Today however, there are more than just a few Charmat (tank) method and carbonated sparkling wines made from British grapes on the market. The joint first Charmat wines on the market (in 2018) were from Flint Vineyard in Suffolk who produced a 'Charmat Rosé' made from 'Solaris, Reichensteiner, Cabernet Cortis and Rondo' which sold well at £22–£24 a bottle. Seven years later they are still making Charmat wines, both white and rosé, both selling at £22.99. The white is made from Bacchus, Pinot noir, Sauvignon blanc, and Solaris and the rosé from Cabernet Cortis, Pinot noir, Reichensteiner, Rondo, and Solaris – presumably in that order of volume. Fitz, based in Worthing, also produced a tank-method sparkling wine in 2019, a white from Chardonnay, Seyval blanc, Reichensteiner, and Alzey 7672 (aka Madeline x Angevine 7672) and a rosé from the same varieties plus Pinot noir and Rondo, both selling for around £18–£20. Five years later they are also still producing the same two wines, so one has to assume they are selling successfully. The Uncommon', who started carbonating wine and putting it into 250 ml cans around 5 years ago, appear to have prospered and their wines now have a wide distribution. They retail at around £6 a can, an equivalent of £18 per 75 cl bottle. Chapel Down are still offering their 'A Touch of Sparkle' carbonated Bacchus which retails at £15 from the winery and around £12.99 in the supermarkets (although I have seen it as low as £9.99 when on offer). I do not like it especially and the one-star reviews on Waitrose's website (20 per cent of the reviews) are not very encouraging. But it has been around for 5+ years now, so it sells.

The big daddy of the world of tank-method English sparkling wine is Mark Dixon. His *Silverhand Estate* now encompasses 365-ha (902-acres) of vineyards making it the second biggest by area in GB. His major vineyards are at Luddesdown just to the south of Gravesend (and where the winery is currently based), in Bodiam, East Sussex, in Sandhurst, Kent and a newly-planted 38-ha one in the Crouch Valley, Essex. They also own Sedlescombe Vineyard, GB's oldest organic producer, and did own Kingscote Vineyard, outside of East Grinstead in East Sussex, but this has now been sold. The original idea, according to Dixon, was to produce an organic tank-method sparkling wine retailing at £12.99. Things have slightly changed since those early days, and the main offering today is their *Silver Reign* tank-method sparkling wine selling from their website at £14.45 a bottle. The white version, made from Chardonnay, Pinot noir and Meunier does not appear to be organic, although the 100 per cent Pinot noir rosé appears to be and is priced at £18 a bottle. They have a number of other wines, still and sparkling, available from their website including what they call their 'flagship' wine, the regally presented *Kyng*, a traditional-method sparkling rosé made 100 per cent from Pinot noir. It will set you back a mere £245 a bottle. However, if you want the special packaging, made out of wood and designed by Lord Norman Foster (which also apparently doubles up as a wine cooler) you will need to find another £150.[12] Their website says that this wine 'has a total release of only 50 bottles'. A collector's item indeed.

Since these early days of non-traditional sparkling wines, the spectrum of different styles, both still and sparkling and of wines made from different grape varieties has proliferated. I attempted to put together a list of the producers with this type of wine but gave up there were so many. There are also orange wines mainly still, pét-nats of all types and grapes, Col Fondo (a cloudy bottle-aged prosecco style) and wines that I will describe as 'natural' for want of a better term. There are also a few non-wines such as vermouth and fruit flavoured wines, plus several producers making distilled products. Given the plantings of

12 Lord Foster has also designed them a £30 mn winery called the Kentish Wine Vaults which is struggling to get planning permission.

both unusual *vinifera* varieties and the multitude of untried PIWIs that are already in the ground (and which will surely not stop expanding now) and given the explosion of different methods of turning grapes into a marketable product and the myriad of styles and types of wine, non-wine and spirits, there is probably almost no variety that cannot be made into something! There is even one vineyard which harvests its grapes unripe and makes a traditional 'verjuice' from them.

How sustainable these non-traditional sparkling wines are is open to question. For producers without vineyards, and therefore dependent on buying their grapes from others, much will depend upon yields, availability and prices. With yields like those in 2018, 2019 and 2023, and with the large number of vines planted in the last ten years, maybe these producers are just what the market needs? Soaking up the excess supplies of grapes and knocking out wines in the £10–£15 price bracket. How profitable a £15 sparkling wine is, is another matter. Deduct 20 per cent VAT and around £3 Excise Duty and you are left with £9.50. Assuming the grapes cost at least £1.50 a bottle, and the production costs are around £5, then the producer is left with £3. If this is sold direct, then it at least shows a return, but if you market via a wholesaler, retailer, or the on-trade (pub, club, hotel or restaurant) then there is very little over to show for your troubles. However, if traditional-method sparklers start selling at £15–£18, is this same price sustainable for a non-traditional-method sparkler? Time will tell. Where these wines will sell well, I am sure, is at the farm-gate where customers can taste them and at events, functions and in the on-trade where wine by the glass is sold without the customer ever seeing the bottle. Asked 'would you like a glass of sparkling wine' most people are not going to ask 'where's it from?' let alone 'what is the method of production?' assuming that they even know the differences between carbonated, Charmat and traditional methods.

Despite the misgivings of some producers and commentators, who view these non-traditional-method wines as somehow debasing 'proper' sparkling wines, it would seem, judging by the number now being made, that they are here to stay. Their future will surely depend on the availability of grapes at the right price, and how low prices on the high street for traditional-method wines will fall. Once Chapel Down starts selling its NV Reserve Brut at £12.99 (a price they are on record as saying they can support) then why buy a Charmat method wine at the same price?

Vine varieties for less favourable regions and cooler sites

With vineyards now being planted all over the world in regions and on sites which a few decades ago were considered too marginal to support commercial viticulture, who can say what 'less favourable' sites are? At the start of the variety recommendations section, I wrote:

All my recommendations are for varieties suitable for COOL climates like Great Britain's. This means climates with:

- An average maximum temperature of +/- 15°C
- GDDs of between 900 and 1,100
- Between 1 and 15 days in the year on average when the temperature reaches 30°C or more
- Climates where temperatures seldom fall below -10°C

Therefore, this section refers to regions, both in GB and in other cool parts of the world, which fall outside these parameters. Of course, there may be mitigations which could improve your site and these need taking into account when selecting a variety to plant. Mitigations such as:

- Partial or complete covering of vines with plastic or fine mesh nets
- Frost prevention techniques
- Protecting vines in the winter months by covering or burying
- Making certain styles of wine or other grape-based product

Wales and Midlands & North regions	Major varieties
Variety	**Ha -2024**
Pinot noir	48.65
Solaris	44.68
Seyval blanc	34.07
Rondo	27.90
Chardonnay	22.19
Bacchus	15.08
Phoenix	12.13
Meunier	10.75
Regent	10.36
Alzey 7672 (Mad x Ang 7672)	7.82
Ortega	5.71
Orion	4.97
Reichensteiner	4.49
Früburgunder, Blauer	3.96
Divico	3.87
Cabaret noir	3.69
Muscaris	2.47
Cabernet Cortis	2.18
Siegerrebe	2.18
Huxelrebe	2.05
Pinot gris	1.84
Other varieties	16.13
Total Wales, Midlands & North	**287.17**

Source: Wine Team at the Food Standards Agency

Table 10

Having started my own vineyard in 1977, when the variety spectrum available to us was much more limited than it is today, when 28°C was a hot summer day and when we never got natural sugar levels much over 9 per cent, I know the challenges of trying to produce wine in a marginal climate. As I have already stated, my choice of varieties for my first vineyard was Müller-Thurgau (of course) and Seyal blanc[13] (of course), plus Reichensteiner (an up-and-coming variety) and Gutenborner, a variety that only one other grower had (and still has) and of which one of my professors at Geisenheim thought highly. Today, climate change has dramatically altered the situation and there are many more varieties to choose from, and many more people growing a multitude of different varieties which will in time, yield valuable data to help growers choose varieties. However, not all of GB is as blessed as the drier eastern and southern regions, and there are many growers who, for whatever reason, wish to establish vineyards in the cooler, windier and wetter parts of the country. For growers determined to plant on more challenging sites, which might be at altitudes over 125-150 m above sea level, and/or with poor shelter from prevailing winds, or just in less-proven and/or less-favourable parts of Britain (and other countries), their choice of varieties must be made with these disadvantages in mind.

Today's list of varieties being grown in Wales and the Midlands & North regions is in marked contrast to even five years ago when Alzey 7672 (Madeleine x Angevine 7672) was one of the top regional varieties and you could still find the old hybrids Cascade, Léon Millot, Maréchal Foch, and Triomphe being used. Acolon was also planted, but this has all but disappeared in the regions covered by the table. In their place, some of the older PIWIs have expanded plus some of the newer ones.

Today's most widely planted vine varieties in WineGB's Midlands & North region which comprises Shropshire, Staffordshire, Leicestershire, Rutland and Lincolnshire and any county to the north of these, plus the whole of Wales, can be seen in Table 10. The top ten varieties comprise five *viniferas* and one hybrid and four PIWIs. Together these ten varieties take account of 81 per cent of vineyards in the two regions. In the next named eleven varieties, there are four PIWIs – Orion (an older cross) and four relatively new ones – Divico, Cabaret noir, Muscaris and Cabernet Cortis. Rounding out the list are five more established *viniferas*: Reichensteiner, Frühburgunder (Pinot noir Précoce), Siegerrebe, Huxelrebe, and Pinot gris. As you can see from Table 11, hybrids and PIWIs

13 We won the Gore-Browne Trophy for 'wine of the year' in 1981 with our second vintage, the 1980 Spots Farm Seyval blanc.

Distribution of PIWIs & hybrids in GB regions				
Region	**Hectares planted**	**Viniferas**	**PIWIs & Hybrids**	**% PIWIs & hybrids**
Wales	91.35	41.48	49.87	54.6%
Midlands & North	195.82	95.89	99.93	51.0%
West	377.90	275.42	102.48	27.1%
T&CVA	138.20	127.96	10.24	7.4%
East Anglia	673.01	631.25	41.76	6.2%
Wessex	594.21	566.07	28.14	4.7%
South East	2,510.64	2,405.52	105.12	4.2%
Total	**4,581.13**	**4,143.59**	**437.54**	**9.6%**

Source: Wine Team at the Food Standards Agency

Table 11

take up just over fifty per cent of the vineyard areas in these two regions, far higher than any other regions. East Anglia, the South-East and Wessex are all below 7 per cent. While this is understandable from a climatic viewpoint, it may be detrimental to overall wine quality.

There is no doubt that one of the reasons for the survival of **Seyval blanc** in the top five of varieties being grown in GB for over seventy years is its ability to produce viable crops in years in which other varieties struggle, its relatively good disease resistance, and its ability to hang on and ripen fruit in conditions that would severely test other varieties. **Reichensteiner,** likewise, while not as disease resistant as Seyval blanc, has also survived, mainly because in good years its yields are higher than the average and these compensate for the poor years. Its grapes usually get high sugars, and the neutral wine is adaptable and can be blended into both still and sparkling wines. In the relatively new (and few) vineyards planted in the more northerly parts of Britain, the selection of varieties is divided into three different sections: the earlier ripening *vinifera* varieties; old hybrids; and new interspecific crosses. The earlier ripening varieties are chosen because they (a) ripen early and therefore stand a chance of ripening on a later site and (b) tend to have lower acidity levels which make them more likely to achieve physiological ripeness and thus make wines of at least a basic quality level.

Solaris has shown itself capable of both good yields, good ripeness levels and good wine quality – the three most important attributes for a successful variety and it will undoubtedly soon become the most planted variety not only in the northern and western regions of GB, but also of the Scandinavian and northern European countries and regions that are establishing vineyards. **Rondo**, despite what I say about its failings in terms of wine quality, is the only red variety that is really worth planting if you are in a marginal region. If your site is at the upper end of marginal, then **Regent** would be worth considering, as would **Phoenix.** Whether you should plant the Champagne trio – **Chardonnay, Pinot noir** and **Meunier** – is down to your experience with your region and your site. If you can be convinced by tasting wines made from these varieties that come from your region and that your site quality is at least as good as the one that the wine you are tasting is coming from, then go ahead and plant. Sparkling wine of course is much more adaptable than still wine and time on lees and the right level of dosage can help an otherwise high-acid, lean dry wine.

Of the other candidates for planting in marginal regions, some of the newer PIWIs, especially early ripening ones, are worth considering such as: **Cabaret noir, Cabernet Cortis, Calardis blanc, Calardis Musqué, Calardis Soleil, Fleurtai, Muscaris, Ravel blanc,** and **Riesel.** Full details can be found in Chapter 9.

Chapter 3

Performance of vine varieties

Yields in Great Britain

There are of course many different factors that affect yield in any region, let alone a marginal region like GB's where weather events, such as a spring frost or a cold and wet flowering period, can badly affect yields. Take 2012 when poor flowering weather caused widespread crop loss. The official national yield was 5.98 hl-ha or around 0.90 tonnes-ha (364 kg-acre). Similarly, 2024, when much of the growing season was unusually wet and cold, the national yield was 21.26 hl-ha, approx. 3.27 tonnes-ha (1.32 tonnes-acre). Compare these with years when there was no spring frost and good flowering weather such as 2018 and 2023 when yields were 45.97 hl-ha and 44.24 hl-ha (6.68 t-ha and 6.43 t-ha) respectively, around a seven and a half times difference. Weather, therefore, is an important factor in yields in marginal regions.

Before weather of course comes climate, and what effect that has upon yields. The climate governs the spectrum of varieties that can be grown, and although GB has seen climate change radically influence our choice of varieties, it is still not warm enough to consider growing high-yielding, hot country varieties such as Chenin blanc or Syrah where in well irrigated sites in say South Africa and Australia, yields of 50 tonnes-ha are achievable. With GDD levels between 850 and 1,050, depending on site and year, the best

2016-2024 Average Yields - all varieties, all regions - Tonnes per hectare

Vineyards 4 years or older	2016	2017	2018	2019	2020	2021	2022	2023	2024	Average t-ha	Average t-acre
Top 25% of vineyards	8.8	9.6	10.5	9.6	7.6	9.1	8.8	13.9	5.9	9.3	3.77
Middle 50% of vineyards	4.4	4.1	6.2	5.3	3.6	5.2	4.6	8.9	2.9	5.0	2.03
Bottom 25% of vineyards	1.4	1.3	1.6	1.8	0.9	2.2	1.4	4.2	0.7	1.7	0.69
All varieties, all vineyards	**4.5**	**4.7**	**7.1**	**5.9**	**4.2**	**5.4**	**5.0**	**9.6**	**4.2**	**5.6**	**2.28**

Source: Data collected by S. P. Skelton from growers (2016-17) and data collected for the WineGB annual yield report (2018-24)

Table 12

GB - Yields of major vine varieties 2016-2024 - all regions, all vineyards

Tonnes per Ha	2016	2017	2018	2019	2020	2021	2022	2023	2024	Average 2016-24	Average tonnes-acre
Bacchus	3.07	4.21	5.48	7.14	3.48	5.91	6.70	9.92	6.20	5.79	2.34
Chardonnay	5.58	4.42	8.67	6.61	6.33	4.75	5.54	9.83	4.30	6.23	2.52
Meunier	5.88	5.24	9.53	6.23	5.36	4.66	6.56	10.49	3.50	6.38	2.58
Pinot noir	3.00	4.31	7.93	6.18	5.25	5.70	6.01	9.20	3.90	5.72	2.31
Averages	**4.38**	**4.55**	**7.90**	**6.54**	**5.10**	**5.26**	**6.20**	**9.86**	**4.48**	**6.03**	**2.44**

Source: Data collected by S. P. Skelton from growers (2016-17) and data collected for the WineGB annual yield report (2018-24)

Table 13

2016-2024 GB	Average Yields - PIWIs and Hybrids all regions										
Tonnes per Ha	**2016**	**2017**	**2018**	**2019**	**2020**	**2021**	**2022**	**2023**	**2024**	**Average 2016-24**	**Average tonnes-acre**
Regent	No data	5.70	11.09	4.12	3.10	6.25	8.09	9.94	2.50	6.35	2.57
Rondo	No data	9.24	7.30	2.66	5.17	6.09	7.01	9.18	2.60	6.16	2.49
Seyval blanc	6.40	11.24	9.38	3.36	4.61	7.39	6.73	9.13	2.50	6.75	2.73
Solaris	No data	No data	No data	2.66	1.99	4.39	4.90	6.45	1.70	3.68	1.49
Averages	**6.40**	**8.73**	**9.26**	**3.20**	**3.72**	**6.03**	**6.68**	**8.68**	**2.33**	**5.73**	**2.32**

Table 14

we can hope for in GB is average yields of around 10 tonnes-ha for sparkling wines and slightly less for good quality still wines. As can be seen from Table 12, over the 9 years 2016-2024, the top quartile by yield of all vineyards and all varieties averaged 9.3 tonnes-ha and only in two years did the yields exceed 10 t-ha. And this is for the very top growers. The middle 50 per cent only averaged half this yield, and the bottom quartile sadly less than a fifth of the top performers.

As can be seen in Table 13 the yield of the four most widely grown *viniferas*, Chardonnay, Bacchus, Meunier and Pinot noir, was 6.03 tonnes-ha, whereas the average for Regent, Rondo and Seyval blanc was 6.42 tonnes-ha, a difference, but a small one. The best performing variety of all was Reichensteiner (See Table 16), a variety that has been grown in GB since the late 1960s, with 7.41 tonnes-ha, admittedly helped by a record shattering 16.60 tonnes-ha in 2018. One also has to remember that Reichensteiner is only grown on 65-ha, so data is by definition quite sparse.

Performance of PIWIs

As can be seen in Table 14, fairly reliable data for three PIWIs, Regent, Rondo, and Solaris, and one hybrid, Seyval blanc, is available. For Regent, Rondo and Seyval blanc, average yields were fairly consistent with them all yielding over 6 tonnes-ha and Seyval just edging the others with an additional 0.50 t-ha. Interestingly, both Rondo and Seyval blanc gave their highest yields in frost-affected 2017, showing the ability of these two varieties to recover after a spring frost. Solaris, for which only six years' data is available, fared less well in all years and ended up with an average yield of only 3.68 t-ha, some 45 per cent less than the other three varieties. Whether this a reflection on the growers, who might for instance have sprayed less, or because they are mainly grown in the less favourable regions of GB, is not known. Compared to most of the *viniferas*, the PIWIs performed comparatively well in terms of yields. Without the 2018 yield, the average yield would be 6.10 tonnes-ha, much the same as the major *viniferas.*

Yields in regions

Yields by Region	2016-24 t-ha	2016-24 t-acre
South East	6.29	2.55
East	6.10	2.47
Thames and Chilterns	5.52	2.23
Wessex	5.39	2.18
West	5.02	2.03
Midlands & North	4.34	1.76
All regions, all varieties	**5.45**	**2.20**

Source: WineGB Yield Surveys of vineyards 0.10 ha or larger & 4 years & older

Table 15

GB - Yields of major vine varieties 2016-2024

Tonnes per active Ha	**2016**	**2017**	**2018**	**2019**	**2020**	**2021**	**2022**	**2023**	**2024**	**Average 2016-24**	**Average tonnes-acre**
Bacchus	3.07	4.21	5.48	7.14	3.48	5.91	6.70	9.92	6.20	5.79	2.34
Chardonnay	5.58	4.42	8.67	6.61	6.33	4.75	5.54	9.83	4.30	6.23	2.52
Früburgunder, Blauer	1.81	3.50	6.62	4.32	3.55	5.34	5.59	6.97	1.80	4.39	1.78
Alzey 7672	6.05	5.64	6.70	6.44	5.22	5.05	8.28	7.17	3.50	6.01	2.43
Meunier	5.88	5.24	9.53	6.23	5.36	4.66	6.56	10.49	3.50	6.38	2.58
Ortega	No data	No data	No data	6.53	3.38	3.75	6.30	8.41	3.30	5.28	2.14
Pinot blanc No data 4.86			7.85	No data	No data	No data	8.88	7.60	3.10	6.46	2.61
Pinot gris	No data	No data	No data	2.21	4.02	4.82	5.26	10.46	3.50	5.04	2.04
Pinot noir	3.00	4.31	7.93	6.18	5.25	5.70	6.01	9.20	3.90	5.72	2.31
Regent	No data	5.70	11.09	4.12	3.10	6.25	8.09	9.94	2.50	6.35	2.57
Reichensteiner	8.36	4.37	16.60	7.50	2.34	N/A	8.20	8.70	3.20	7.41	3.00
Rondo	No data	9.24	7.30	2.66	5.17	6.09	7.01	9.18	2.60	6.16	2.49
Seyval blanc	6.40	11.24	9.38	3.36	4.61	7.39	6.73	9.13	2.50	6.75	2.73
Solaris	No data	No data	No data	2.66	1.99	4.39	4.90	6.45	1.70	3.68	1.49
Averages	**5.02**	**5.70**	**8.83**	**5.07**	**4.14**	**5.34**	**6.72**	**8.82**	**3.26**	**5.83**	**2.36**

Source: Data collected by S. P. Skelton from growers (2016-17) and data collected for the WineGB annual yield report (2018-24)

Table 16

As might be expected, yields in the warmer and drier eastern half of GB are higher than the windier and wetter western half, although there are many factors to consider. Site selection, vine density, vineyard management and choice of grape variety are all critical and it would be good to be able to get data subdivided by some of these factors. Unfortunately, this is not currently possible. However, the data collected over nine years has hopefully ironed out some of the anomalies and Table 15 shows clearly the east-west yield differences. The average of the three eastern regions is almost 20 per cent higher at 5.97 tonnes-ha (2.42 tonnes-acre) whereas the yield for the three western regions is 4.92 tonnes-ha (1.99 tonnes-acre). Wales was excluded from this table as there was just not enough reliable data.

Potential alcohol levels

With reference to Tables 17 and 18 one can see that the potential alcohol levels of all the major varieties plus Alzey 7672 and Orion. Of course, picking dates are governed by many factors, wine style perhaps being a major factor. Still or sparkling, white, red or rosé – all may need different ripeness levels. The four major varieties, Bacchus, Chardonnay, Meunier and Pinot noir, were all remarkably consistent with their average potential alcohol levels between 9.1 and 9.3 grammes per litre. Sugar levels of the three major PIWIs and the hybrid Seyval blanc were mixed. Regent and Rondo averaged 9.30 per cent and 9.00 per cent ABV respectively, straddling the all-variety average of 9.2 per cent ABV, but with Solaris, with its good disease resistance and early ripening, averaging 10.4 per cent ABV, a significant 1.2 per cent ABV advantage. Despite the fact that most of the Solaris is growing in the cooler regions of GB, for it to come up with this performance is impressive. It does show that PIWIs can perform well and in theory ought to

GB - White vine varieties potential alcohol levels at harvest 2016-2024

Variety	2016	2017	2018	2019	2020	2021	2022	2023	2024	Average 2016-24
Alzey 7672	9.8	8.6	9.4	8.7	9.5	9.0	9.5	8.9	9.5	**9.2**
Bacchus	9.5	9.1	9.6	8.5	9.4	8.7	9.5	8.6	8.9	**9.1**
Chardonnay	9.7	9.4	9.8	8.4	9.6	8.3	10.2	8.7	8.8	**9.2**
Orion							8.6			**8.6**
Phoenix	9.0	8.1					8.8		8.0	**8.5**
Pinot gris	11.8	9.3			8.8		9.7	8.4	9.9	**9.7**
Reichensteiner	10.2	10.0	9.7	9.5	9.4		8.7	9.8	9.7	**9.6**
Seyval Blanc	8.3	7.6	8.6	7.8	9.1	8.0	9.4	7.6	8.0	**8.3**
Solaris	10.8	10.9		9.8	9.3	10.4	11.2	10.1	11.0	**10.4**
Average for year	**9.9**	**9.1**	**9.4**	**8.8**	**9.3**	**8.9**	**9.5**	**8.9**	**9.2**	**9.2**

Source: Data collected by S. P. Skelton from growers (2016-17) and data collected for the WineGB annual yield report (2018-24)

Table 17

GB - Red vine varieties potential alcohol levels at harvest 2016-2024

Variety	2016	2017	2018	2019	2020	2021	2022	2023	2024	Average 2016-24
Frühburgunder, Blauer	11.7	10.2	9.2	10.0	10.2	9.9	9.9	9.4	10.2	**10.1**
Meunier	10.0	8.7	9.4	8.5	9.5	8.6	9.4	8.5	8.8	**9.0**
Pinot noir	9.9	9.1	9.8	8.6	9.6	8.7	9.9	8.6	9.1	**9.3**
Regent		10.0	9.6	7.9	9.0	9.1	10.2	8.7	10.0	**9.3**
Rondo (GM 6494/5)	9.5	9.6	8.7	8.6	6.7	9.0	9.9	8.9	10.5	**9.0**
Average for year	**10.3**	**9.5**	**9.3**	**8.7**	**9.0**	**9.1**	**9.8**	**8.8**	**9.7**	**9.3**

Source: Data collected by S. P. Skelton from growers (2016-17) and data collected for the WineGB annual yield report (2018-24)

Table 18

be capable of producing wines with above average quality. Seyval blanc, which has been a stalwart of the industry since it started in 1953, is well known for having low natural sugar levels at harvest, and this is shown in the data. At an average of 8.3 per cent ABV, this is the lowest average level recorded, although well above the 5 per cent and 6 per cent levels that we were used to in the '70s and '80s. Of course, for sparkling wines which (like all wines) can be enriched by 3 per cent (or 3.5 per cent in poor years) and then gain an additional 1.5 per cent during the secondary fermentation, a low starting sugar level is not so much of a problem.

Acidity levels

Tables 19 and 20 show the acid levels at harvest, measured in grammes per litre as tartaric acidity.[14] As with sugar levels, the differences between mainly still and mainly sparkling

14 Most GB wineries work in grammes per litre as tartaric. A few (mainly French inspired ones) work in sulphuric. To convert tartaric to sulphuric multiply by 0.65 i.e. 10 g/l as tartaric become 6.5 g/l as sulphuric.

GB - White wine varieties - Acidity at harvest (g/l tartaric)

Colour	Type	Variety	2016	2017	2018	2019	2020	2021	2022	2023	2024	Average 2016-24
white	Vinifera	Alzey 7672	9.0	8.6	10.1	9.5	9.8	8.8	10.1	8.0	8.8	**9.2**
white	Vinifera	Bacchus	9.6	9.2	9.0	9.2	8.5	9.4	8.9	8.1	9.3	**9.0**
white	Vinifera	Chardonnay	14.1	14.0	12.0	12.9	11.7	15.0	11.9	10.8	14.7	**13.0**
white	PIWI	Orion							11.9			**11.9**
white	PIWI	Phoenix		10.3					9.4		11.1	**10.3**
white	Vinifera	Pinot gris (Ruländer)	9.8	10.9		11.1	7.6		10.6	9.9	11.1	**10.1**
white	Vinifera	Reichensteiner	9.0	8.8	8.8	9.3	10.2		8.2	8.5	9.3	**9.0**
white	hybrid	Seyval Blanc	13.0	11.5	11.3	11.3	10.3	11.2	11.9	10.5	11.1	**11.3**
white	PIWI	Solaris	8.6	9.5		8.3	8.0	8.1	8.8	8.1	10.1	**8.7**
		Average for year	**10.4**	**10.3**	**10.2**	**10.2**	**9.4**	**10.5**	**10.2**	**9.1**	**10.7**	**10.3**

Source: Data collected by S. P. Skelton from growers (2016-17) and data collected for the WineGB annual yield report (2018-24)

Table 19

GB Red vine varieties - Acidity at harvest (g/l tartaric)

Colour	Type	Variety	2016	2017	2018	2019	2020	2021	2022	2023	2024	Average 2016-24
red	Vinifera	Frühburgunder, Blauer	5.9	8.7	8.5	8.5	9.5	8.4	8.3	8.5	8.3	**8.3**
red	Vinifera	Meunier		13.3	11.9	12.7	10.7	12.9	12.2	10.3	13.8	**12.2**
red	Vinifera	Pinot noir	13.0	12.2	12.0	12.3	10.7	12.8	11.3	10.3	13.3	**12.0**
red	PIWI	Regent		10.5		9.4	7.5	9.6	8.2	8.1	9.1	**8.9**
red	PIWI	Rondo (GM 6494/5)		10.3	9.4	10.7	9.0	10.2	9.6	8.4	10.8	**9.8**
		Average for year	**9.4**	**11.0**	**10.4**	**10.7**	**9.5**	**10.8**	**9.9**	**9.1**	**11.1**	**10.2**

Source: Data collected by S. P. Skelton from growers (2016-17) and data collected for the WineGB annual yield report (2018-24)

Table 20

varieties are quite marked. The highest acid levels are recorded for Chardonnay (13 g/l), Meunier (12.2 g/l) and Pinot noir (12 g/l), the three main sparkling wine varieties. The fourth highest (ignoring Orion for which there is only one data point) is for Seyval blanc at 11.3 g/l which is again a 'mainly sparkling' variety. The PIWIs with good data, Regent, Rondo and Solaris, all recorded lower than average acidity levels, 8.9 g/l, 9.8 g/l and 8.7 g/l probably reflecting their use as mainly for still wines where higher sugars and lower acids are required.

Chapter 4
Vine varieties – the legal situation

Yotes Court Vineyard, Kent. The nicest site I ever planted.

Until Britain reached a total of 500-ha (1,235-acres) of planted vines, MAFF[15] was reluctant to get involved with the actual vine varieties growers had planted in their vineyards. Although in theory growers were only meant to plant those varieties on the 'Recommended' and 'Authorised' lists, little was done to enforce this. These lists, which had been drawn up prior to Britain's entry into the Common Market in 1973, were subsequently updated by MAFF's 'Vine Varieties Classification Committee' and brought into line with the varieties actually in the ground. Once a planted area of 500-ha was reached, a fact MAFF reluctantly discovered in the 1987 voluntary vineyard census, growers in Britain were restricted to planting varieties on the lists. However, as long as Britain's wine production was under 5,000 hl in any one year, there were no restrictions on the area that could be planted and there would be no planting ban, something that existed in all other Member States with sizeable vineyard areas. However, as the area under vine in Britain increased and with it, volumes produced, the limit on wine production before a planting ban would be introduced was increased to 25,000-hl average annual production over 5 years and upped again to a 50,000-hl average annual production over 5 years. Since then, the area under vine has grown massively and today (2024) the 5-year average production level stands at 93,315-hl.

Having left the EU (on 31 January 2020 at 11.00) and adopted *en masse* the EU wine regulations, progress in changing things has been slow. Discussions are under way with DEFRA about the way forward on everything to do with viticulture and wine production and no doubt one day there will be a revised set of rules and regulations. As far as vine varieties and planting vines are concerned, it is probably very safe to say that a planting ban is not likely to happen in the foreseeable future and most probably never. Having said that, given the huge (and, of course, uncontrolled and currently uncontrollable) increase in planting in recent years, some existing growers have suggested that a brake on planting might not be such a bad thing, but this is pure speculation. If changes to the ability to plant vines freely were ever to be mooted, DEFRA would require a substantial majority of votes in favour of it (not just a simple majority) and most probably it would be on an area planted basis, not one grower, one vote. Therefore, a few large-scale growers who would be against a ban could easily swing the vote. Certainly, for the time being, a vine planting ban in Britain can be forgotten.

As far as which varieties may or may not be planted, DEFRA and Wine Standards appear to be fairly relaxed about it at the moment. The EU

15 Ministry of Agriculture, Fisheries and Food, today DEFRA – Department for Environment, Food & Rural Affairs.

rules (the ones we adopted in 2020) stated that six old American and/or hybrid varieties – Clinton, Herbemont, Isabella, Jacquez, Noah, and Othello[16] – were not permitted to be planted anywhere in the EU and I have no reason to think that this has changed (not that you would want to plant any of them in GB). Apart from these six, there are no restrictions on the vine varieties you can plant and no restrictions on the area you can plant. Any variety may be used for winemaking, subject to the winemaking and labelling regulations for the category of wine you propose to make (and hopefully sell – I say 'hopefully' as you might make a PDO wine which fails the tests and/or tasting and has to be sold under a lower category label). These regulations govern when you are allowed to use a variety name or names and what other descriptors you may use and they are different for each category of wine. PDO sparkling wines can only be made from certain varieties which are set out in the regulations. The lists of acceptable varieties in the current regulations are out of date and have hybrid varieties listed as acceptable for PDO wines (which they are currently not), as well as other errors. If in doubt, ask Wine Standards.

Categories of wine that may be produced in England and Wales

Following changes to the basic EU wine legislation in 2011, there are now three distinct categories of wine which can be produced in Britain:

- Wines with 'Protected Designation of Origin' (PDO), also called Quality Wines. Testing and tasting required.
- Wines with 'Protected Geographical Indication' (PGI), also called Regional Wines and which are table wines with some additional quality parameters. Testing and tasting required.
- 'Wine' which is any wine that is not PDO or PGI. No testing or tasting required.

The last category used to be known as 'varietal wines' which had to go through a 'Certification' process, but from 1 January 2024 this has not been required. All PDO and PGI wines have to be both analysed and tasted and must pass the laid-down criteria before they may be so labelled.

Until these changes took place in 2011, only still wines made in Britain were eligible to be labelled as Quality Wines. To gain Quality Wine status, wines had to be made from *vinifera* varieties, i.e. no hybrids; had to be made to certain quality parameters; and all wines had to be tested and tasted. This system is basically still current. Somewhat confusingly, sparkling wines made to certain standards could (and still can) be called Quality Sparkling Wines even though they can be made from any variety, including hybrids. This is because the term 'Quality' in 'Quality Sparkling Wine' refers to the method of sparkling, not the base wine from which the *cuvée* is made.

After the 2011 changes, Britain gained the right to make PDO Quality Sparkling Wines which are the sparkling equivalent of Britain's still Quality Wines. The most important difference between PDO Quality Sparkling Wines and non-PDO Quality Sparkling Wines is that the PDO wines can only be made from certain varieties. In theory, only grapes grown in vineyards below 200 m above sea level may be made into PDO wines, but whether this is enforced is a different matter. To use the term 'quality' on the label, sparkling wines must be made by the traditional-method. Only sparkling wines made by the traditional-method may use this term i.e. not carbonated or tank method. There are also yields restriction for both PDO and PGI wines. These are a maximum of 80 hl-ha for PDO wines and 100 hl-ha for PGO wines. Sussex PDO wines, still and sparkling, may not exceed 12 tonnes-ha 'under normal conditions' with an 'absolute maximum' of 14 tonnes-ha. However, 'dispensation may be given ... in exceptional years' to these yield limits. Quite what 'normal' and 'exceptional' mean are not set out in the regulations.

16 As I write this, I read in the Times that the European Parliament has just voted to allow these six varieties to be released from their prison and planted again. Given their wine quality however, I doubt whether many hectares will actually be planted.

PDO, PGI and other categories of wines

There are currently (2025) four separate PDO regions: England, Wales, Sussex (comprising the counties of East and West Sussex), and Darnibole which is a Camel Valley Vineyards *monopole.* All applications for PDO and PGI status are dealt with by a 'Wine Scheme Manager' who is part of WineGB's team.

PDO Still Wines

English Quality Wine, Welsh Quality Wine, Darnibole Quality Wine, Sussex Quality Wine[17].

PDO Sparkling Wines

English Quality Sparkling Wine, Welsh Quality Sparkling Wine, Sussex Quality Sparkling Wine.

PGI Still Wines

English Regional Wine, Welsh Regional Wine

PGI Sparkling Wines

English Regional Quality Sparkling Wine, Welsh Regional Quality Sparkling Wine

Non-PDO and PGI Still and Sparkling Wines

Allowed to state origin, e.g. Kent and can say 'Wine of England' or 'Wine of Wales', but use of the protected term 'English' or 'Welsh' or 'English Regional' or 'Welsh Regional' or 'Sussex' is not permitted on labels. They may also bear vintage and variety name(s) subject to varietal labelling rules.

17 The label descriptors are: 'Sussex', 'Sussex Sparkling', 'Sussex Still', and 'Sussex Origin'.

Chapter 5

Vine variety descriptions – *Vinifera* varieties

The twenty *vinifera* varieties described in this chapter are the most widely grown in Britain in 2024. Together they amount to 4,136.24-ha (10,220.6-acres) which is 89 per cent of the total planted area of 4,650.9-ha. The varieties listed in the chapter on PIWIs and hybrids occupy 439.09-ha, which together with the viniferas in this chapter account for 99.24 per cent of all varieties planted in GB vineyards.

Major vinifera vine varieties 2024

Variety	Ha	Variety	Ha
Acolon	5.75	Meunier	410.78
Albarino	4.47	Müller-Thurgau	37.43
Alzey 7672	36.62	Ortega	45.17
Auxerrois	7.25	Pinot blanc	53.85
Bacchus	333.81	Pinot gris (Ruländer)	79.94
Chardonnay	1,498.28	Pinot noir	1,369.77
Dornfelder	15.85	Reichensteiner	64.53
Frühburgunder, Blauer	95.00	Sauvignon blanc	56.17
Gamay noir	7.69	Schönburger	19.92
Huxelrebe	15.62	Siegerrebe	18.34

Source: Wine Team at the Food Standards Agency

Table 21

Notes to variety descriptions:

- **Synonyms:** In general, I have used French variety names wherever possible as these are the names most likely to be found on bottles of wine in Britain and in vineyard literature. Thus, you will find references to Pinot rather than Burgunder or Spätburgunder; Meunier rather than Schwarzriesling or Müllerrebe; Pinot gris rather than Pinot Grigio, Ruländer, or Grauburgunder; although I have used Blauer Frühburgunder, rather than Pinot noir Précoce, as this is a German variety.
- **Madeleine x Angevine 7672:** Following the investigations and DNA profiling of a variety which we have called various things over the years, what I have referred to as Madeleine x Angevine 7672 has now been officially renamed Alzey 7672.
- **Meunier:** I have used Meunier, the correct name for the variety often called Pinot Meunier (or even sometimes Wrotham Pinot).
- **Müller-Thurgau:** When referring to Müller-Thurgau in the individual variety descriptions I use the abbreviation MT.
- **Sylvaner:** I have kept to the old spelling of Sylvaner when referring to both 'Madeleine x Sylvaner III 28/51' and 'Riesling Sylvaner' (Müller-Thurgau or Rivaner) in preference to the modern German spelling, 'Silvaner', except where this variety was one of the parents of another variety. The Swiss (and others) continue to use the spelling Sylvaner.
- **Databases:** There are two different European vine databases which don't necessarily agree with each other. The JKI database is I believe the more accurate and up to date of the two.
- **JKI Database:** *Vitis* International Variety Catalogue VIVC www.vivc.de
- **VIVC:** This reference is to the European *Vitis* Database www.eu-vitis.de

Vine-breeding establishments

References are made in the variety descriptions to the following vine-breeding establishments:

- Alzey, Rheinpfalz, Germany
- Blattner, Valentin, Soyhières, Switzerland
- Freiburg, Staatliches Weinbauinstitut, Germany
- Geisenheim Institute for Grapevine Breeding (Gm), Rheingau, Germany
- HBLA Klosterneuburg, Austria
- Julius Kühn Institute (JKI) for Grapevine

Breeding, Geilweilerhof, Germany[18]

- Mendel University, Brno, Czech Republic
- Oppenheim, Rheinhessen, Germany
- Potapenko All-Russia Research Institute, Rostov-on-Don, Russia
- Vivai Cooperativi Rauscedo (VCR), Italy
- Weinsberg, Baden-Württemberg, Germany
- Würzburg, Bayern, Germany

Sugar and acid levels

Sugar levels are usually given in per cent potential alcohol or sometimes ABV (alcohol by volume) which is typically the label declaration level. Most British growers and winemakers use degrees Oechsle (°OE) when discussing ripeness levels at harvest, although some growers use °Brix as they have refractometers which only show this calibration.

Acid levels in GB are usually expressed in grammes per litre of total acidity as tartaric, as they are in almost all other countries, except for France and French-influenced regions where levels of total acidity as sulphuric acid are used. To convert from one to the other multiply tartaric by 0.66 or multiply sulphuric by 1.52. Therefore 6 g/l as sulphuric becomes 9.12 g/l as tartaric.

PIWI and hybrid vine varieties for vineyards in Great Britain

For the following varieties please see Chapter 9

Blütenmuskateller	Fleurtai	Rinót
Cabaret noir	Johanniter	Rondo
Cabernet blanc	Muscaris	Sauvignac
Cabernet Cortis	Orion	Sauvignon Rytos
Cabertin	Phoenix	Seyval blanc
Cal 1-28	Pinotin	Solaris
Calardis blanc	Prior	Soreli
Calardis Musqué	Ravel blanc	Souvignier gris
Calardis Soleil	Regent	Voltis
Donauriesling	Réselle	
Divico	Riesel	

Acolon

Type: *Vinifera*
Colour: Red
Origin: Weinsberg, Germany
Crossing: Limberger x Dornfelder
Date of crossing: 1971
Year of registration: 2002
Planted area in Germany (2024): 422-ha
Planted area in GB (2024): 5.75-ha

Acolon is a German *vinifera*-cross bred in 1971 at Weinsberg by Bernd Hill. Its parents – Blauer Limberger (also known as Blaufränkisch in Austria) and Dornfelder – are both staples of German red wine production. Dornfelder itself is Helfensteiner (Blauer Frühburgunder x Black Hamburg) crossed with Heroldrebe (Portugieser x Limberger), so its antecedents are impeccable. In Germany it was released in 2000 and grew to a total of 461-ha by 2018. Since then the area has reduced and is now down to 422-ha (2024). In

18 Many of the newer Geilweilerhof varieties have the prefix 'Calardis' which is an old Franconian name for the village where the JKI is based.

Germany it is planted because of its very dark red wine which adds colour to Dornfelder wines of which there are plenty and it has higher sugars, although a slightly lower yield than Dornfelder. However, since global warming occurred, better red wine varieties have come to take its place – Cabernet Sauvignon and Merlot for example.

In Britain it first appeared on the variety lists in 2009, New Hall Vineyards being the first to plant, and by 2015 had risen to 7.45-ha. Since then it has reduced to 5.75-ha and it is no longer considered as a creditable option for red wine production. As the climate has improved, Pinot noir, with its heritage, class and consumer acceptability, will continue to be the red variety of choice. In addition, there are several red PIWIs which are better and more disease resistant.

Albariño

Type: *Vinifera*
Colour: White
Origin: Original variety
Planted area in GB (2024): 4.47-ha

Albariño – or Alvarinho if you are Portuguese – is a staple variety of the north-west of the Iberian Peninsula and found both as a single varietal and in blends throughout this area. It is one of the major varieties found in Vinho Verde. It is a variety that has been 'discovered' in recent decades and is now grown in many different parts of the world. At its best it is Riesling-like, with crisp acidity and good length, an ideal wine to accompany food, especially fish and shellfish.

In Britain it first appeared on the variety lists in 2013 and has now grown to 4.47-ha (2024) and is a variety that could, with a bit of encouragement from the climate, become more widely grown. Sandhurst Vineyards in Kent have been growing 0.40-ha (1-acre) since 2012 and have achieved some success with it. They made a Bacchus-Albariño blend which drank very well and a wine made from 100 per cent Albariño in 2014 surprised many Spanish wine judges when it was tasted at a blind tasting in Madrid. The judges described it as 'full, rich and artisanal' which is hopefully a compliment in Spain. Chapel Down, who are major buyers of Sandhurst grapes, have signed up to take all their Albariño. Its limitations for growers in Britain are a low yield, and late-ripening and high-acid grapes, but on a good, warm site and with patience, it might be worth trying. This is definitely a variety to watch out for.

Alzey 7672

Synonyms: Madeleine x Angevine 7672, Madeleine Angevine, Mad Angie
Type: *Vinifera*
Colour: White
Origin: Alzey, Germany
Crossing: Madeleine Angevine x Riesling
Date of crossing: 1944-45
Year of registration: Not known
Planted area in GB (2024): 36.62-ha

In 1957 Ray (Barrington) Brock[19] at Oxted was sent cuttings of several varieties by Dr Zimmerman from Alzey, and one of them was labelled 'Sämling [seedling] 7672'. The crossing was made in the 1944-45 season when Georg Scheu – responsible for varieties such as Huxelrebe, Faberrebe, Kanzler, Regner, Scheurebe, Septimer, Siegerrebe and Würzer – was the Alzey Institute's director. By 1960 Brock was able to report that it was "giving large crops which ripen with Riesling Sylvaner [MT]. Considered to be a promising variety". Not having been given the crossing details of Sämling 7672 Brock wrote to Zimmerman and asked for them. He was informed that it was a 'freely pollinated seedling of Madeleine Angevine'. When Brock started to sell cuttings of the variety (as he did with all promising varieties) he gave it the name Madeleine Angevine 7672 and it was under this name that it was known for many years. As it

19 Ray Barrington Brock (1908-1999) was one of the founders of the modern wine industry in GB and in 1945 established what came to be known as the Oxted Viticultural Research Station at his home in Oxted, Surrey. Here he trialled over 600 different varieties and published several books about his exploits.

became quite popular, the name on wine labels became shortened to simply 'Madeleine Angevine'. This was unfortunate as another variety already existed under this name.

The true Madeleine Angevine (also sometimes called Madeleine d'Angevine), is a female only table grape variety, for many decades said to be a crossing of Précoce de Malingre and Madeleine Royale,[20] made by Pierre Vibert at the Moreau-Robert nurseries in Angers in 1857 (some reports say 1859), but described in *Wine Grapes*[21] to be a crossing of a Chasselas-based table grape variety called Circé and Madeleine Royale. The VIVC database gives it as a crossing of Blanc d'Ambré and Madeleine Royale. It is one of the earliest table grape varieties for open cultivation in France and named after St. Madeleine's Day (22 July), said to be the date on which it can first be harvested. Having only female flowers and being very early, it has been used by plant breeders in a number of crosses over the years. Morio and Husfeld used it for Forta and Noblessa and it is one grandparent of Reichensteiner. Scheu himself used it to produce Siegerrebe and, interestingly, this variety, which at one time was credited with being a Madeleine Angevine x Gewürztraminer crossing, was unmasked by Heinz Scheu, Georg's son, as also being a freely pollinated Madeleine Angevine seedling. It is probable that the variety we now grow in Britain as Alzey 7672 comes from the same crossing programme that produced Siegerrebe.

In 1992, I asked Professor Alleweldt (or Professor Dr. Dr. h.c. Diplomlandwirt Gerhardt Erich Alleweldt to give him his full title), then Director of the Geilweilerhof State Institute for Grapevine Breeding, to see if he could discover more about our Mad Angie (as it is often known). He located Georg Scheu's old breeding books for the 1944 - 45 season in the Alzey archives and found that 'Sämling 7672' was – as Brock was originally told – an 'open pollinated progeny of Madeleine Angevine'. In 2019, I was asked by Geilweilerhof to send leaf samples of 'our' Madeleine Angevine for DNA profiling. Three samples were sent from different vineyards in Britain and, a few weeks later, Erica Maul from Geilweilerhof emailed to say that two of the samples[22] were actually a crossing of Madeleine Angevine x Riesling and that henceforth it should be known as Alzey 7672 and this is how it would be listed on the VIVC database. Thus, after 75 years, the mystery of Mad Angie's birth parents was cleared up. (There is also another variety to cause further confusion called Madeleine Angevine Oberlin, a crossing between Madeleine Angevine and Bouquettraube.)

Alzey 7672 (as we should now call it) found favour with many of the early British growers and it became fairly widely planted and at one time it was the third most popular variety in GB (after MT and Seyval blanc). Gillian Pearkes, one of the pioneers of British grapegrowing, who was an influential character in the West Country, thought highly of it and many of the early pioneers planted it. Unfortunately, owing to the confusion over the name and the fact that this variety was not available from any other source than Brock's original stock or vineyards planted with vines obtained from Brock, some growers were sold vines of the true table grape variety Madeleine Angevine, which was barely suitable for Britain except in the very best years. Unless it was being grown in proximity to other early varieties for pollination, it seldom set a good crop and, in some years, ripened at the end of August. The wines from this variety were flabby and never really acceptable except for blending. The confusion over this variety – foreseen in a rare moment of sanity (as far as rules and regulations governing British viticulture is concerned) by the European Commission in 1973 when MAFF submitted vine varieties for classification – has

20 Madeleine Royale, a table grape vine variety once thought to be a Chasselas seedling, but now shown to be a Pinot x Black Hamburg cross, has been 'outed' by the gene-jockeys as the father of Müller-Thurgau, with Riesling being the mother.

21 *Wine Grapes* by Jancis Robinson MW, Julia Harding MW and JoséVouillamoz, Allen Lane, London, 2012.

22 The third sample turned out to be Auxerrois!

meant that it gained something of a chequered reputation.

Alzey 7672 is easy to grow, ripens early, gives good crops and the wines have a light muscat tone and low acidity. In 1988 I took wood from what were known to be genuine Alzey 7672 growing in Robin Don's Elmham Park vineyard in Norfolk, now grubbed up, which Robin had planted with cuttings supplied by Pearkes (who had got hers directly from Brock). Buds from this wood were then sent to France and grafted on to resistant rootstock. The resultant grafted vines were sold to various growers around the country, including Sharpham Vineyard in Devon who made several award-winning wines with it. Sadly their vineyards were grubbed in around 2000. Alzey 7672 reached its peak at the end of the 1990s with 68.70-ha, but since then it has slowly declined and today (2024) stands at 36.62-ha.

Alzey 7672 is still being planted by a few growers in the further reaches of Britain's vinegrowing regions as its low acidity means that it will ripen at lower heat levels than many of our mainstream varieties, although where the bud wood comes from is anyone's guess. For the longer term, however, I suspect that what we colloquially call Mad Angie will not survive the Pinot-Chardonnay storm-troopers lining up against her and as her growers get older and give up, sell up or go to that vineyard in the sky, the area will dwindle. Pity, as the variety can make good wine. For growers with more challenging sites, it is probably a variety worth considering, but they are advised to make sure they know where the bud-wood comes from before buying any vines.

Auxerrois

Synonym: Pinot Auxerrois
Type: *Vinifera*
Colour: White
Origin: Original variety
Planted area in Germany (2024): 310-ha
Planted area in GB (2024): 7.25-ha

Auxerrois, an ancient variety, related to Chardonnay and Pinot blanc, is widely grown in Alsace where it is usually blended into *Edelzwicker* (although in Alsace it can also be labelled as Chardonnay or Pinot blanc) or used for sparkling wine. It can also be found in Burgundy and Luxembourg in the Old World, and Canada, New Zealand and the USA in the New World. Geisenheim has produced some productive clones and there are over 300-ha planted in Germany. It was included on the original list of 'Recommended' varieties submitted to Brussels when Britain joined the Common Market in 1973, even though it was not then widely grown and had not been suggested by the EVA.

Auxerrois is a variety of moderate vigour, fairly resistant to *botrytis* and ripens its wood well. It is quite a late variety, ripening in Britain after MT but before Seyval blanc. As a neutral Pinot blanc/Chardonnay style variety which would be useful for barrel ageing or as a sparkling wine base, it should be more widely planted in Britain – perhaps on the more challenging sites – as it has lower acid levels than other similar varieties. It is a steady yielding variety and it is a pity that it appears to have fallen out of favour in Britain, but I guess that if you can ripen Chardonnay, why would you want to grow Auxerrois. There were 9.00-ha in 2009, but only 7.25-ha in 2024, so it is not getting any more popular. However, I believe that it is 'one to watch' and well worth a trial.

Bacchus

Type: *Vinifera*
Colour: White
Origin: Geilweilerhof, Germany
Crossing: (Silvaner x Riesling) x Müller-Thurgau
Date of crossing: 1933
Year of registration: 1972
Planted area in Germany (2024): 1,515-ha
Planted area in GB (2024): 333.81-ha
(See table 13, page 28.)

Bacchus, a crossing made by Peter Morio and Professor Husfeld at Geilweilerhof in 1933, was first registered in 1972 and was known in the Rheinpfalz as the 'Early Scheurebe'. Its parentage is the same as Optima. Bacchus is the second most popular white *neuzüchtung* (new-crossing) in Germany (after Kerner) and in 2024 there were 1,515-ha, although considerably down from the 1995 area of 3,449-ha. With global warming, German growers are moving away from their home-grown fruity, sweet wine varieties (Bacchus in Germany is almost always made in a *Spätlese* or even an *Auslese* style) towards more international, food-friendly varieties such as Chardonnay and the Pinots.

Bacchus first appeared in Britain in the Wye College vineyard in 1973 as an experimental variety and was upgraded to 'Recommended' in 1998[23]. Bacchus, as a variety for Britain, seems to be here to stay and the area under cultivation has risen from 76-ha in 1990, via 295-ha in 2020 to 333.81-ha in 2024. While this is not exactly rocket-propelled growth, it is at least growth. Bacchus is one of the best white varieties being grown in Britain today for the production of still wines and it regularly wins many of the major prizes. In growth habit it is similar to MT, although perhaps not quite as vigorous, but just as prone to *botrytis*. It appears to ripen readily in most vineyards, although in cooler years, acids can be high and care needs to be taken that they are not too high in the bottle, especially with wine made from grapes harvested at lower sugar levels. In years with early springs and high summer temperatures, Bacchus can suffer from sunburn and an early de-leafing is advised in order to get the grapes used to the sunshine.

In the bottle, Bacchus falls into two camps: what one might term the *Sauvignon de Touraine* or pretend *Sancerre* camp, where the wine is light, fruity, with a modest spiciness, but of no great weight; and the full-on Marlborough Sauvignon style with luscious, spicy, even catty, fruit with enough body and weight to carry the flavour. These tend to be the riper (higher natural sugar) examples and are best when they are bottled with perhaps a few grammes of residual sweetness (as are many Marlborough wines). Chapel Down's Kit's Coty Bacchus is often a fine example of this latter (and to me, more preferable) style. This wine has won the 'best Bacchus' trophy six times in the last ten years. In the 2017 Decanter World Wine Awards (DWWA), the 2015 Winbirri Bacchus won a Platinum Award and 'Best Single Varietal Wine under £15' which got huge press coverage and introduced the variety to a whole new audience.

With top Bacchus wines selling for £15–£20 a bottle, this is a variety that can make money. The one *really* good thing about Bacchus is the name – free from the umlauts and unfortunate prejudices associated with Germanic names – just a name everyone recognises and thinks has something to do with wine. Who was that Bacchus fellow anyway, some sort of God?

Chardonnay

Type: *Vinifera*
Colour: White
Origin: Original variety
Planted area in Germany (2024): 3,050
Planted area in GB (2024): 1,498.28-ha
(See table 13, page 28.)

With its origins lost in the mists of viticultural time, Chardonnay is now found in virtually every grapegrowing region in the world, from the hottest parts of Australia, South Africa and California to one of the coolest of all growing regions, Britain. With its being the dominant (often the only) variety in white wines from Burgundy and, of course, in most Champagnes, it is not surprising that early vineyard owners in Britain were seduced into thinking that it would do well here. Brock had it in his collection at Oxted but could never get it to ripen properly. Salisbury-Jones planted it at Hambledon in the late 1950s and had the same problems: excessively high acid levels and low natural sugars. Ian and Andrew Paget at Chilsdown Vineyard (Singleton, West Sussex) planted Chardonnay and I seem to recall one year (1981?) when the acidity (in grammes per litre as tartaric) was higher than the

23 See Appendix IV or the story of the Wye College experimental vineyard.

degrees Oechsle! Ouch. Extreme unripeness was a common finding among those early growers who persevered with it, although most decided to give up and removed the offending variety. Only in really hot years would it produce anything like ripe grapes and tolerable wine.

The first vineyard I can find that grew Chardonnay and produced some drinkable wine is New Hall. They inadvertently planted a few bundles of the variety which had been mistakenly included in some Pinot gris they had ordered. Piers Greenwood, son of the New Hall founders Bill and Sheila Greenwood, told me that in 1984 they picked 1,300 kgs of Pinot noir and 150 kgs of Chardonnay, 'the very last grapes picked from the harvest', and sent them to Lamberhurst Vineyards to be made into sparkling wine. This is the wine which Chris Trembath, Piers' brother-in-law and now part-owner of New Hall Vineyards, remembers having at his wedding in 1987[24]. In the late 1980s, more growers started to plant Chardonnay – New Hall planted more, plus Throwley, Surrenden and of course, Nyetimber – for the production of bottle-fermented sparkling wines.

In my 2001 book *The Wines of Britain and Ireland* I wrote:

through a combination of good site selection and traditional training systems (aided by a degree of global warming) they have started to produce some interesting results. While acids are still high at harvest (15 g/l is not uncommon) the combination of a full malolactic fermentation and the traditional secondary bottle fermentation, renders them manageable.

Well, the 'interesting results' turned out to be more than interesting and since then some extremely good wines, many sparkling, but more and more very respectable still wines, have been produced. High acids can be a problem and growers, especially those with less than perfect sites, need to make sure they select the right clones and rootstocks, plus prune and canopy manage, to get acids down. Chardonnay plantings in GB have gone from 7.0-ha in 1988, via 33.5-ha in 2002, 119.6-ha in 2007, 327.0-ha in 2013, 975-ha in 2020 to a (seemingly) massive 1,498.28-ha in 2024, thus making it Britain's most widely planted variety and accounting for just over 32 per cent of the planted area. Given the way the British climate seems to be heading I can only see Chardonnay going from strength to strength. The quality of the best *blanc de blancs* sparkling wines can be superb and they age extremely slowly and extremely gracefully. In recent years and in warm vintages, the still wines made from Chardonnay can be good, and I firmly believe now that in time we will be producing the Chablis, and the Chablisiens had better get planting pineapples. I hear that some well-known Burgundy producers have been seen scouting around vineyard sites in Essex, so watch this space!

As a variety for Britain, Chardonnay appears to be more and more at home. Viticulturally it is no more demanding to grow than most varieties. It buds up quite early and in frost-prone sites this can be a problem. It is susceptible to Powdery and (especially) Downy Mildew and *botrytis* and it almost always needs spraying in its first year, i.e. the year of planting. Until it can be demonstrated otherwise, simple *Guyot* single- or two-cane pruning appears to work well. While we might like to have them, we do not get the very high yields common in Champagne (15–20 tonnes-ha is not uncommon), so I do not think cordon pruning is required, although I know some growers are trying it. It benefits from an open canopy and deleafing both straight after flowering and pre-*véraison*. It will also get late-season Downy Mildew, which will affect the ripening process, so protection needs to be kept going until the harvest interval starts.

Chardonnay ripens late and is usually not harvested until the third or even fourth week of October. I did hear of one grower who picked a 'late harvest' Chardonnay on 28 November which must be the latest British harvesting date ever. Of course, 'late' is subjective and having rarely

24 See Appendix 1 for more details on this wine.

enjoyed sharing Guy Fawkes night (5 November) with my children in the late 1970s and 1980s as I was always pressing Seyval blanc, today's Chardonnay growers shouldn't complain too much. Potential alcohol levels are usually in the 8–10 per cent region – ideal for sparkling wine – although in cool years and late sites 7-8 per cent is quite common. In really warm years – 2018, 2023 and 2025 being examples – natural alcohol levels can reach 12–13 per cent, figures that, forty years ago, Chablis struggled to achieve. I have heard that in 2025, one Crouch valley grower harvested his Chardonnay at more than 15 per cent per cent natural alcohol. An amazing figure for GB.

Acid levels at picking will depend very much on whether the grapes are for still or sparkling wine. For sparkling wine, Chardonnay is typically picked at 12-15 g/l although in cool years and on late sites, 16–17 g/l is quite common and even higher acidity levels are not unheard of[25]. For still wine, depending on the style of wine required, 10-12 g/l should be what growers are aiming for, but in poor vintages and from cool sites, these lower figures may not be possible. However, with de-acidification, malolactic fermentation, some blending with lower-acid varieties, extended *sur latte* ageing and an appropriate *dosage*, even high-acid wines can turn out fine. It just shows what time and patience will achieve. The real trick is to get the level of malic acid down.

For sparkling wines made in Britain, Chardonnay is an indispensable component and no serious grower can be without it. I could select Chardonnay-based *blanc de blancs* wines from any one of the top growers – Chapel Down, Furleigh, Gusbourne, Harrow & Hope, Nyetimber, Ridgeview, Sugrue South Downs, Wiston – and know that I would be getting a great wine.

For the production of still wines, Chardonnay has shown that with a suitable site, correct canopy management and, most importantly, the right level of crop, very good wines can be produced in Britain. Chapel Down, Blackbook Winery, Danbury, Gusbourne, Simpsons and several other producers have produced some very good still wines from Chardonnay, especially from ripe vintages. If growers can consistently make still wines that sell at the same sort of prices as middling Chablis and white Burgundy, then we will see even more Chardonnay planted for still wine in years and decades to come. It will be interesting to see what Jackson Family Wines come up with from their Essex vineyards.

The combination of a classic name, its Champagne heritage and many excellent sparkling (and increasing number of very good still) wines, have made Chardonnay indispensable for any vineyard in Britain with pretensions to seriousness. I am not convinced that it suits all the sites it has been planted on in the last two decades, and some growers may yet struggle to ripen it and/or find that their wines need years (5-plus) in the bottle to come round – not a recipe that I think will guarantee financial success.

Note: See the separate section towards the end of this chapter on: *Clones of Chardonnay, Pinot noir and Meunier*

Dornfelder

Type: *Vinifera*
Colour: Red
Origin: Weinsberg, Germany
Crossing: Helfensteiner x Heroldrebe
Date of crossing: 1955
Year of registration: 1980
Planted area in Germany (2024): 6,507-ha
Planted area in GB (2024): 15.85-ha

Dornfelder is one of oldest of the 'new' German crossings, bred in 1955 at Weinsberg in Württemberg, in the heart of Germany's red wine producing region, Here, the traditional varieties – Limberger, Trollinger (Black Hamburg) and

25 One very well-known early sparkling wine producer once confided in me that their maiden vintage *blanc de blancs* had an acidity at picking of 20 g/l as tartaric. It took considerable efforts to get a malolactic going, plus over 4 years *sur latte* to tame the acidity. However, it did go on to win a major award.

Schwarzriesling (Meunier) – suffer from a lack of colour and substance and Dornfelder was bred to produce more of both. The crossing was made in 1955 by August Herold and released to growers in 1980. Dornfelder's parents are two other Weinsberg varieties: Helfensteiner (which is Blauer Frühburgunder x Black Hamburg) and Heroldrebe (which is Portugieser x Limberger). Dornfelder is the most widely grown *neuzüchtungen* in Germany, and its 6,507 ha (2024), although down from what was probably its peak, 8,197-ha in 2012, still occupy 20 per cent of their red vine area and 6.4 per cent of their total vine area. It is capable of producing some good wines, albeit in a spicy Rhône style, rather than a classic Bordeaux style, although some German examples can be fairly light and insubstantial, probably due in part to high yields: 120–150 hl-ha is not unknown.

Dornfelder first appeared in Britain in the late 1980s and is capable of producing tolerable red wines, especially in blends with other varieties. It is no more difficult to grow than other *vinifera* varieties and in good years can produce quite heavy crops. Very few 100 per cent Dornfelder wines have been produced in Britain and it is mainly blended with other reds. It can certainly make good rosés in blends. Plantings expanded from 5.40-ha in 1990 to 20.00-ha in 2015 but have been declining ever since. I suspect that it will continue to decline and suffer a slow lingering death. For true reds, Pinot noir, Blauer Frühburgunder (Pinot noir Précoce), Regent and Rondo are all preferred and there are a number of new PIWI varieties which are probably better choices. My advice: no need to rip it out but think twice before planting any more.

Frühburgunder, Blauer

Synonym: Pinot noir Précoce
Type: *Vinifera*
Colour: Red
Origin: Old selection
Planted area in Germany (2024): 212-ha
Planted area in GB (2024): 95.00-ha

Blauer Frühburgunder, more often known in GB by its French synonym Pinot noir Précoce, is a relatively old variety, a mutation of Pinot noir, first planted in the 1800s in the Ahr wine region, which was, until reunification, Germany's most northerly winegrowing region[26]. The region is named after a tributary of the Rhine which joins the main river just south of Bonn. By the 1900s the variety had migrated to the town of Ingelheim am Rhein, '*Die Rotweinstadt*' (the red wine town) and it became that town's dominant variety, and the wines became very well known[27]. Geisenheim has done some work on it, cleaned it up (the old clones had Leaf Roll and cropped very poorly) and today there are five clones listed for growing in Germany although here it is reducing in area from 262-ha in 2012 to 212-ha in 2024.

In Britain, it has been grown since around 1990. When growers started planting Pinot noir in earnest in Britain, the search went out for the earliest ripening clones in the belief that Spätburgunder, as it is generally known in Germany, was a late ripening variety (it's not, but that's another matter) and they were offered what many thought was an early German clone of Pinot noir by some British-based vine suppliers, but was, in fact, what the Germans call Frühburgunder. Vines were imported and planted and the wine was sometimes called by its correct name, (Blauer) Frühburgunder, but was more usually called Spätburgunder or Pinot noir, or occasionally Early Pinot noir. Everything was fine until some

26The most northerly German wine region today is Saale-Unstrut, and the most northerly vineyard is in the village of Eisleben, birthplace of Martin Luther.

27 Ingelheim, situated on the southern side of the Rhine, about halfway between Mainz and Bingen, is where Charlemagne built a palace (in about 800). According to legend, he noticed that the snows melted first on the south-facing slopes opposite and could see that the area had a unique microclimate. His son, Ludwig the Pious, started growing grapes there and the slopes eventually became the site for what is today Schloss Johannisberg.

eagle-eyed WS inspector noticed Frühburgunder on a bottle label and realised that this variety was not on the list of permitted varieties for Britain and could therefore only be sold as UK Table Wine. After some discussion with the industry, the WS agreed that the synonym 'Pinot noir Précoce' would be a permitted name (as it is already called this in other EU member states and is a name that is so much better than Blauer Frühburgunder). Growers with this variety were asked to confess their misdeeds and re-register their vineyard parcels containing this 'clone'. This is why it suddenly appeared from nowhere on the variety lists. Frühburgunder ripens up to two weeks earlier than Pinot noir and on good sites achieves better sugars and better colour. However, it is just as *botrytis*-prone as straightforward Pinot noir and offers no other viticultural advantages. I used to be somewhat negative about the variety, suggesting that it was only fit for growers whose sites were not good enough for 'proper' Pinot noir, but since then, no doubt helped by climate change and some good vintages, my opinion of it has changed and I believe it has a future in Britain. Growers also seem to share this view, and the area under vine has gone from 19.94-ha in 2011 to 95.00-ha in 2024. If you are wanting to make good, still red and rosé wines, and have a good site, then the variety is worth trying. Where I believe it should not be planted is on poor sites and in areas where Pinot noir will not ripen well.

Gamay noir

Type: *Vinifera*
Colour: Red
Origin: Original variety
Planted area in GB (2024): 7.69-ha
Planted area in France (2020): approx. 25,000-ha

Gamay is one of the oldest varieties grown for wine production and was first recorded in the late 1300s in France. It was a variety liked by vignerons because of its high yield, but its quality was never as good as Pinot noir, the other major variety grown in the same regions, and it was regularly banned from cultivation in the more up-market appellations. Its best-known home is of course Beaujolais where it is the only red variety allowed and often made into wine using carbonic maceration which produces wines with a lot of fruit character, but not much substance. However, when cropped moderately and treated correctly, it can make fine wines.

In Britain, Gamay also has a long history and it is more than likely that in the Middle Ages, many monastic vineyards were planted with a version of it. When the third Marquess of Bute sent his gardener to France in 1873 to look for varieties to plant at his Castle Coch site outside Cardiff, it was Gamay noir that he ordered (along with a white variety called *Millie Blanche*) and planted in the spring of 1875. The Bute vineyards expanded over the next 35 years, mainly with cuttings taken from the original 0.15-ha (0.37-acres). These vineyards eventually went out of production during the First World War and were grubbed up in 1920. It is recorded that in the four decades of their existence, the grapes only ripened in seven years!

In more modern times, Gamay has been grown in GB since the late 1970s and Biddenden Vineyards in Kent have used if for both still and sparkling wines. It has never shone however as a high-quality variety and even today, with global warming help provide it with a warmer environment, it has seldom (if ever) made wines of much distinction. Having said that, in the 2025 WineGB competition the only still wine from the 2024 vintage to win a gold medal was the Cobham House Rosé made from Gamay. In recent years, a few more growers have planted small areas and the area has grown from 0.63-ha in 2018, to today's area of 7.69-ha. I personally would not recommend it for growing anywhere in GB, let alone on a prime warm site which is what it needs to make anything decent, but it is being planted a bit more, hence the entry in these pages.

Huxelrebe

Type: *Vinifera*
Colour: White
Origin: Alzey, Germany
Crossing: Elbling x Courtillier Musqué
Date of crossing: 1927
Year of registration: 1968
Planted area in Germany (2024): 311-ha
Planted area in GB (2024): 15.62-ha

Huxelrebe is a crossing made by Georg Scheu at Alzey in 1927 and is named after Fritz Huxel, a grower from near Worms who first recognised its potential. Originally said to be Chasselas x Courtillier Musqué (also known as Muscat Précoce de Saumur), DNA analysis has now shown it to be Elbling x Courtillier Musqué. In Germany, where the area planted has been falling steadily since its heyday in the 1990s and now stands at 311-ha in 2024, it is capable of producing very large yields of grapes (or as the Germans say *'trägt wie ein Esel'* – 'hung like a donkey'), with high natural sugars.

Huxel – as it is usually called – was first introduced to Britain via the Wye College experimental vineyard in 1972 and on account of its higher yielding ability was quite widely planted. In 1999 there were 59.1-ha, but since then the area has fallen and is today down to around 15.62-ha, and I doubt if anyone is planting it today. Huxel can be extremely vigorous, with great fat canes and very large leaves and is probably best grown on a spur-pruned system such as Geneva Double Curtain (GDC) where vigour can, to a certain extent, be tamed. Bunches are very large and contain an unusually high number of seedless grapes (*millerandage*) which ripen readily and achieve very high natural sugars, although unfortunately loved by wasps. It undoubtedly benefits from bunch thinning and pre-harvest deleafing as it suffers badly from *botrytis*. When really ripe, the wine can be very fruity with a pronounced Sauvignon blanc character, although acid levels can be high and, to obtain the best wines, the grapes must be allowed to ripen fully. Unripe examples can be rather too herbaceous and catty and can be detected even when disguised in a blend. It has been used for a few late-harvest sweet wines, but only when *botrytis* has got the better of it. In 2001 I wrote:

a variety that is probably worth persevering with on account of its good quality wine and higher than average yields. With a new generation of anti-botrytis fungicides now available, will Huxelrebe become more viable to grow?

Well, almost twenty-five years on, I can answer the question: 'No'. It is still a pushy monster in the vineyard, over-cropping when it feels like it and suffering from wasps and rot in equal measure. Its diminishing area sums up its attractiveness to British growers, although organic grower Will Davenport likes it and uses it for his 'Hux' wine[28] and it has made a number of award-winning wines. According to WS records, there has been none planted since 2018.

Madeleine x Angevine 7672

See entry for Alzey 7672

28 The best British-grown Huxelrebe I ever tasted was the late Bill Ash's 1982 Staple St James Huxelrebe, one of the very few English wines I have ever bought a case of. The interesting story about this wine is that Bill sold half the grapes to Lamberhurst (he was always strapped for cash) and they made a wine with it, sweetened with some high-strength *Auslese* Gewürztraminer *süss-reserve* that Karl-Heinz Johner had prized out of the local co-op winery where he lived in Baden. This 1982 Staple St James Huxelrebe won the 1983 Gore-Browne Trophy, largely on account of the great *süss* used, but fell apart after a while and the Staple wine with, as I recall12 per cent natural alcohol, aged into a really great wine. The *süss* trick was one I remembered, and when I was winemaker at Lamberhurst I used the same Gewürztraminer *süss* in the 1988 Schönburger which won the Gore-Browne in 1990! They say all's fair in love, war and winemaking. Pilton Manor also produced a very good late harvest Huxel in 1992. This was a complete accident and the crop had been written off, but then owner Jim Dowling was persuaded by his winemaker, John Worontschak, to pick the crop despite it being completely affected by *botrytis*. It was an excellent wine and I recently read a report of a tasting at which it was still showing very well.

Meunier

Synonyms: Pinot Meunier, Schwarzriesling, Müllerrebe – and, in Britain only – Wrotham Pinot
Type: *Vinifera*
Colour: Black
Origin: Original variety
Planted area in Germany: 1,598-ha (2024)
Planted area in GB (2024): 410.78-ha

Meunier, one of the numerous mutations that go to make up the Pinot family, accounts for around one-third of the plantings in Champagne (the actual percentage appears to be a closely guarded secret) and while it is seldom spoken about in the same breath as its nobler companions – Chardonnay and Pinot noir – and is not allowed in *Grand Cru* wines, it is used in some of the finest *cuvées* (Krug for example). In Champagne it is favoured for its higher acidity than Pinot noir, its later bud-burst and its higher yields. It is found in cooler sites and on north-facing slopes, both of which mean that bud-burst is later and therefore less frost-affected. It is characterised by its white tipped shoots and its hairy white leaves, especially on the underside, which give it the 'dusty' look from which the name – which means 'miller' in French – stems. Like many of the Pinots, it can show reversions back to the true Pinot noir and it is quite possible to have vines with both hairy and non-hairy leaves on different shoots in one year, which disappear in another year.

In Britain, Meunier has had a somewhat chequered career. In the early 1950s, Edward Hyams, one of the early pioneers of British grape growing, discovered a vine growing on a cottage wall at Wrotham in Kent which, in a fit of journalistic fever, he named Wrotham Pinot and which he gave to Ray Brock at Oxted. Given that this foundling had the classic dusty and red-edged leaves of Meunier, Brock assumed, wrongly or rightly, that it probably was a Meunier, although he reported that it had 'a higher natural sugar content and ripened two weeks earlier' than supplies of Meunier obtained from overseas. Brock sold cuttings and it became quite popular with some of the early vineyards. It is highly doubtful now whether any cuttings from Hyams or Oxted still survive in vineyards in Britain, and all plantings of Meunier stem from France or Germany (where it is known as Schwarzriesling or Müllerrebe). However, the name Wrotham Pinot is still a permitted synonym for Meunier in Britain.[29] In 1990 there were only 5.5-ha being grown in Britain, most of it fairly old and unloved, but since 2006, when the area stood at 22.5-ha, plantings have increased and today (2024) it is the third most widely planted variety and there are 410.78-ha of it. There is no doubt now that it is seen as an indispensable part of many *classic cuvée* sparkling wines and often accounts for 15–20 per cent of Champagne-variety plantings.

Although it has been grown for over fifty years in British vineyards, until quite recently, Meunier has never really shone as a variety capable of making interesting wines on its own and most of it has been blended with other varieties. However, with rising plantings, there is much more interest in Meunier and my experiences to date, despite my former misgivings about its suitability for our climate, have been favourable. Today, there are more and more wines, both still and sparkling, white, red and rosé, made wholly from Meunier, or where it is the dominant partner. This mini-explosion of Meunier has proved that it is adaptable, and can make good, even very good wines (especially still reds in warm vintages). Table 13 (page 28) shows that in terms of yield, Meunier can more than hold its own with its Champagne cousins, although the data is skewed owing to the 2018 yields. This is because the Meunier bunches were very, very large – sone growers reported them around the 250-300 gramme range – well over double their normal range. Still red wines from Meunier are extremely rare anywhere in the world, and if Britain can show that Meunier is the equal of its more respected cousin, Pinot noir, then it might become our USP.

Note: See separate section at the end of this chapter on: *Clones of Chardonnay, Pinot noir and Meunier*

29 See *Appendix II* for the full story of Wrotham Pinot.

Müller-Thurgau

Synonyms: Rivaner, Riesling Sylvaner
Type: *Vinifera*
Colour: White
Origin: Geisenheim, Germany
Crossing: Riesling x Madeleine Royale[30]
Date of crossing: 1882
Year of registration: 1923
Planted area in Germany (2024): 10,511-ha
Planted area in GB (2024): 37.43-ha

Professor Dr Hermann Müller, a Swiss national from the canton of Thurgau, near Zurich, produced this crossing in 1882 while working at Geisenheim. Returning to Switzerland in 1891 to become Director of the Wädenswil Research Institute, Professor Müller asked for 150 of his best crossings to be sent to him, including No.58, which was eventually to become MT. Owing to some confusion with the labelling at the time the crossings were delivered, the true parentage of crossing No.58 was never discovered and it eventually became known as Riesling Sylvaner[31] on account of its wine style, said to resemble a blend of Riesling and Silvaner. Professor Becker, head of Geisenheim in the 1970s and 1980s, attempted to recreate the variety by making multiple Riesling and Silvaner crossings, but failed. He was, however, of the opinion that it more resembled a Riesling x Riesling crossing than any other.

In 1996 an Austrian, Dr Regner, from the viticulture school in Klosterneuburg, thought that he had proved that the crossing was between Riesling and a member of the Chasselas family, a table grape vine variety which (so he thought) was called Admirable de Courtiller. It was then discovered that the reference vine of this variety in Klosterneuburg's collection was in fact another variety altogether – it turned out to be Madeleine Royale, a variety that was once thought to be a Chasselas seedling, and according to the *Vitis* database is actually an 1845 crossing of Pinot and Black Hamburg.[32] In 2001, two researchers in Germany, Erika Dettweiler and Andreas Jung, were able to unravel the DNA in MT and prove that it was indeed a crossing between Riesling and Madeleine Royale: after 119 years MT's parents were finally found.

In 1912, wet-sugaring (the use of sugar in solution to chaptalise wines) was forbidden in Switzerland and MT started to replace Elbling, up until then a widely grown variety, but high in acid. MT then began to find favour in Germany and, in the early 1920s, was taken up by Georg Scheu, then at Alzey in the Rheinland-Pfalz. He subjected it to clonal selection and helped improve its yield. Following the Second World War, when many vineyards were suffering from *Phylloxera* as well as the ravages of the war, MT was widely planted and its large yields, early ripening and soft wines, lower in acidity than either Riesling or Silvaner, were much appreciated. It was once Germany's most widely planted vine variety, mainstay of all *Liebfraumilch* and by 1999 occupied an area of 20,672-ha which was 20 per cent of the total German vineyard area. However, market forces and global warming have combined and made the variety less popular and today (2024) it is down to 10,511-ha occupying 10.3 per cent of the total German vineyard area.

In Luxembourg – where it is called Rivaner – it is still a popular variety, planted on around 250-ha which is 20 per cent of their total vineyard area of 1,250-ha. It crops very well, 150 hl-ha is not unknown – and is usually blended into both still and sparkling wines. Until 1987, it was New Zealand's most widely planted variety and accounted for 42 per cent of their vineyard area although plantings have now almost disappeared. It is popular in high-altitude vineyards in Italy's German-speaking Alto Adige (Südtirol) region, where it can make some very fine wines. It is also

30 Madeleine Royale is Pinot x Schiava Grossa (Black Hamburg) made in 1845 at the Moreau-Robert Nursery in Angers. (VIVC)

31 Sylvaner with a 'y' is the old spelling, still used in Switzerland. For some reasons the Germans banned its use and stipulated that it be spelt Silvaner. One grower I know rebelled and simply called his wine 'Y'.

32 Black Hamburg is a synonym for Trollinger.

to be found in many other European countries, including Austria, Hungary, Czech Republic, Slovakia and Switzerland. EU regulations forbid the use of the name Riesling Sylvaner, although this name can still be found in non-EU countries such as Switzerland.

Its introduction into Britain stems from Ray Brock's visit to Switzerland in 1946 when he met Mr Leyvraz at the Swiss Federal Vine Testing Station at Caudoz-sur-Pully. In 1947 vines of various varieties were sent by Leyvraz to Brock, including 'Riesling Sylvaner'. Brock gave cuttings to Edward Hyams in 1949 and they both trialled it for a number of years. Brock first harvested grapes from MT on 14 October 1950. In Hyams' 1953 book *Vineyards in England* Brock wrote in his chapter on vine varieties that 'Riesling Sylvaner is known to give an outstandingly fine wine in cool climates'.

When Jack Ward was looking for vines for the Horam Manor vineyard, planted in 1954, MT was recommended to him and it became a staple of the vineyards of the day. Ward was a very influential figure in the early days of the revival (he was the EVA's first Chairman) and undoubtedly did much to persuade growers to plant MT. Ward's company – the Merrydown Wine Company – sold vines and gave advice and ran a winemaking co-operative between 1969 and 1979 to which many of the early growers belonged.[33] MT was top of my list when I selected varieties for planting in 1976–7 and it was almost unthinkable to plant a vineyard in Britain then without it. Most German wine and viticulture experts considered England to be a country of mists[34] and not-very-mellow fruitfulness and thought that a high-cropping, early ripening, low acid variety (which is what MT is in Germany) was just what we needed.

MT is a vigorous variety, especially in its early years, and it will grow thick canes and large leaves, leading to excess shading. This often results in poor cropping, especially in years with low light and heat levels in the previous season. It suffers from *botrytis*, Powdery and Downy Mildew and requires regular spraying, and is best deleafed both after flowering and then again around *véraison*. In heavy-yielding years, stem-rot can be a problem. The wood often ripens poorly and does not overwinter well, often showing active *botrytis* in mild winters. It is probably a variety best avoided by organic growers. Since the advent of better anti-*botrytis* products, disease control has become easier (if more costly) and clean crops of MT are the norm, rather than the exception they were in the 1970s and 1980s.

The grapes of MT can have good fruity flavours with light muscat hints and, at their best, the wines made from them can be very good. However, when less than fully ripe, they tend towards the herbaceous and catty. The acidity is usually average to low and with some balancing residual sugar, the wine can be very attractive and fresh when young and will keep well, although probably best drunk within two to three years after bottling. Picked early with sufficient acidity and when the fruit flavours are more neutral, it can be used for sparkling wine, and Chapel Down always used it in their entry-level *Vintage Brut Reserve*. This has now been replaced with pressings from Chardonnay and Pinot noir.

In 1984, MT was by far the largest single variety being grown in Britain, with 149.0-ha, accounting for 35 per cent of the vineyard area. Over the intervening years, however, it has slowly fallen to 61.5-ha in 2009 and is now down to only 37.43-ha (2024) which is under one per cent of the vineyard area. It would appear that a few growers are hanging on and no doubt, in these warmer times,

33 Ward was an interesting man and spoke fluent German. He was an accomplished pianist having taken a music degree as a young man and had then studied the piano at the *Frankfurt Konservatorium* and learnt the language.

34 When I lived and studied in Germany, people would often remark when I said I came from England *'ahh – sehr neblig'* – very foggy, harking back to all the old Sherlock Holmes films they had seen which depicted London wreathed in pea-soupers. I got so fed up with this that I went to the local weather station in Geisenheim and asked them how many official 'foggy days' they had. It turned out that (being on the river) they actually had more 'foggy days' than London, so I was able to say in reply *'nicht vahr'* – not true – and give them the data.

getting good enough crops to justify its existence. However, no one is planting it anymore and, over time, it will continue to dwindle. With hindsight, it was a variety that Britain could probably have done without, although one must not forget that the most popular wine at the time it was being planted was the light, fruity *Liebfraumilch* style that MT is so suitable for. Its extreme vigour, especially in the early years, coupled with its on-off cropping pattern and disease problems, make it a difficult variety for British conditions.

Ortega

Type: *Vinifera*
Colour: White
Origin: Würzburg
Crossing: Müller-Thurgau x Siegerrebe
Date of crossing: 1948
Year of registration: 1971
Planted area in Germany (2024): 324-ha
Planted area in GB (2024): 45.17-ha

Ortega is a crossing made in 1948 by Dr Hans Breider at Würzburg between MT and Siegerrebe. Siegerrebe is a freely pollinated Madeleine Angevine seedling with Gewürztraminer now confirmed as father. Ortega was first registered in 1971 and is named (somewhat curiously) after the Spanish philosopher José Ortega y Gasset, (but good for pub quizzes when 'what's the name of a famous Spanish philosopher' comes up.)

In Germany there are 324-ha (2024) mostly in Rheinhessen and the Pfalz. However, it is a declining variety there, down from 951-ha in 2001. In Germany, it ripens early, achieves high natural alcohol levels and low acidity levels, and with its rich, spicy tones, makes a good blending partner for higher acid varieties such as Riesling.

In Britain, Ortega was introduced by Jack Ward, who planted it at Horam Manor in 1971, and it has slowly grown in popularity. It ripens early – usually just after Siegerrebe and Optima – has high sugars, low acids and plenty of flavour. An early bud-burst makes it susceptible to spring frost damage and it is sensitive to difficult flowering conditions and will suffer from *coulure* in poor years. It can be quite vigorous and canopy management needs to be good to get the best fruit. It will also get *botrytis* towards the end of ripening which will turn to 'noble rot' if sugar levels are high enough. Given good canopy management and timely attention to spraying, this variety can provide high-quality grapes, useful for both normal still wines and dessert wines. Growers report that its bunches tend to get tangled up with each other and with the canes and wires and picking therefore can take twice as long as for other varieties.

When fully ripe, wines made from Ortega are rich and zesty with good balance, although warm years may result in wines with rather low acidity and care needs to be taken to pick at the correct time. Biddenden Vineyards have won numerous gold and silver medals with their Ortega wines and won the Gore-Browne Trophy with one in 1987. Surprisingly, it takes to new oak well and Chapel Down used to make a very good barrel-aged *Condrieu*-like wine from Ortega. For Westwell Vineyards in Kent it is something of a speciality and their 'Orange Ortega' skin-contact wine wins many awards. Growers of Ortega seem to like the variety and the area has risen from 29.5-ha in 1990 to 45.17-ha in 2024 which must be seen as a small vote in its favour. I suspect that it will remain as one of our important, albeit minor, varieties and, as vineyards creep ever northwards, might find favour there.

Pinot blanc

Synonym: Weißerburgunder, Pinot Bianco
Type: *Vinifera*
Colour: White
Origin: Original variety
Planted area in Germany (2024): 6,362-ha
Planted area in GB (2024): 53.85-ha

Pinot blanc is one of the most widely distributed of varieties across Europe and is one of the large family of Pinots. It is often confused (one suspects mostly on purpose) with Chardonnay – the style of wine they produce can be similar – and in general terms it is less demanding than Chardonnay, will ripen more easily and has a higher yield. In Germany it is increasing in popularity and there were 6,362-ha in 2023 making it the fourth most widely planted white variety. It is to be found in France, in Alsace especially, and in many Italian regions in great quantity (especially in the Alto Adige where it can make superb wines) and in many other temperate regions.

In Britain it was first grown in 1979, but in only one vineyard and did not start to be more widely planted until 2005-06 when global warming enlightened us to its possibilities. Since then it has grown steadily from 6.2-ha in 2006 to 53.85-ha in 2024 and looks as if it will keep being planted. Like Chardonnay, it requires a good site and careful management to ripen it fully and get the acids down, although it is no more difficult to grow than Chardonnay. Pinot blanc bunches tend to be very large and more compact than Chardonnay, and in order to avoid *botrytis* it needs timely deleafing and spraying.

Initially it was used for the production of sparkling base wine, but in the last twenty years quite a few 100 per cent varietal still wines have been made and been awarded many prizes. Chapel Down's, made from grapes grown in Hampshire, Essex and Kent, is one of their best still wines and more growers should follow their example. Acids can be at the higher end of the spectrum, but by allowing the grapes to ripen fully, and then putting the wine through lees-ageing, malolactic fermentation, and maybe a couple of years in bottle, some very good still wines can be made. As a useful variety that will find a good home in both sparkling wine blends and high-end still wines it deserves to be planted more. It has a good pedigree, an acceptable name, and is gaining a track record. I expect the area under this variety will continue to expand.

Pinot gris

Synonym: Grauburgunder, Ruländer, Pinot Grigio,
Type: *Vinifera*
Colour: White
Origin: Original variety
Planted area in Germany (2024): 8,404-ha
Planted area in GB (2024): 79.94-ha

Pinot gris is another variety from the large Pinot family and it appears to have almost as many clones as there are winegrowing regions that use it. In Germany there were 8,404-ha in 2023, up from 7,069-ha in 2019, making it the third most widely planted white vine variety, and the area under vine has more than trebled since 1995 and will surely overtake MT in the not too distant future. In Germany it is known as Ruländer for wines with residual sugar and Grauer Burgunder or Pinot gris for soft, fruity dry wines. In Alsace, its most respected home (where until a decade or two ago it was known as Tokay d'Alsace), it is capable of producing both good to very good dry and off-dry wines, as well as exceptionally fine late-harvest dessert wines – *Vendange Tardives* and *Sélection de Grains Noble*. In Italy, where there are said to be over 30,000-ha (the exact figure is hard to obtain), it makes a very neutral, quite crisp wine, ideal with food. It seems to be able to change its style to suit the region. In the new world it can be found in many countries. New Zealand, now that it has stopped making Pinot gris wines with residual sugar, can produce very good examples. There are numerous clones of Pinot gris. VCR, the biggest vine nursery in Italy, has around fifteen different ones and there are three listed for France.

In Britain, where it has been grown since the late 1970s, it is making good progress from 6.0-ha in 2007, to today's total of almost 80.0-ha. This makes it one of the fastest growing varieties, faster even than Chardonnay and Pinot noir, albeit on a much smaller scale. It is not the easiest variety to ripen and requires a good site and careful canopy management to make the best of it. Yields are not the highest and need to be kept under control to produce the best grapes. Some reports talk of a prolonged flowering, resulting in picking in two or more passes. There are probably some clones which would suit our conditions more than others, but (again) who is doing any trials? As plantings have increased over the last decade or so, the number of good to very good wines made from it has also increased, helped a bit by warm vintages such as 2018, 2023 and 2025. It has also been used in sparkling wines, but, except in a blend, I cannot see it becoming an important variety for this type of wine.

Assuming that the climate continues to improve, and as more single varietal wines from Pinot gris become available and achieve success, I can see it making steady progress and becoming a minor, but useful, variety for producers in Britain.

Pinot noir

Synonym: Blauer Spätburgunder, Spätburgunder, Pinot Nero
Type: *Vinifera*
Colour: Red
Origin: Original variety
Planted area in Germany (2024): 11,437-ha
Planted area in GB (2024): 1,369.77-ha
(See table 13, page 28.)

Pinot noir is one of the most ancient of varieties and probably has more clones and variants than any other variety. While its home is often thought of as Burgundy, where undoubtedly many of the finest examples can be found, it seems to thrive in both very warm and very cool climates and good examples can be found in Australia, California, Chile, Germany, New Zealand, South Africa, Spain and even Norway. In Germany, the planted area has gone from 7,199-ha in 1995 to 11,437-ha in 2024 (which is about the same as it was in 2022) but it is still the most widely planted red vine variety, occupying almost 36 per cent of the red wine area and 11.1 per cent of the total vine area. It is also, of course, not only used for still red wines, but is also one of the classic sparkling wine grapes, found not only in Champagne, but wherever good sparkling wines are made.

In Britain it was one of the earliest varieties to be grown. Brock trialled it at Oxted, but was not pleased with it and by 1961, when he issued *Report No.3 Progress with Vines and Wines,* he stated that Pinot noir 'appears to be very much later [than on the continent] in this climate... and has been discontinued'. His lack of success was probably due to the fact that, at the vineyard's elevation of 125–137 metres above sea level, the site was just too cool. In addition, Brock was not, at least not in the early years, looking to make sparkling wines, although he did make some later on. Jack Ward was likewise somewhat dismissive of the variety and also thought it ripened too late.

Despite the reservations of these pioneers, it seems to have been planted quite widely, if not in any great quantity. Many of the early growers seemed to limit their investigation of continental vineyards to a quick visit to Champagne and returned enthused with the idea that Pinot noir, and for that matter Chardonnay, were naturals for our climate. The truth of the matter is that in those early days they were not, and good wines made from them were *very* few in number. Whatever Bernard Theobald, one of England's most flamboyant viticultural pioneers, claimed in the 1970s and 1980s, Pinot noir was *not* the best variety to be growing then. What I wrote in 2001 (I hope) bears repeating:

The rise in interest of bottle-fermented sparkling wines and the results to date, from those vineyards that have gone into the job seriously, is impressive. There have also been a number of very creditable red wines made from the variety in recent years – only it has to be said in the warmer years – but it is an exciting trend. There are a number of factors which might make Pinot noir a more acceptable variety for Britain. There is no doubt that vine breeders have done much to change Pinot over the last 50 years. Clones are available now that ripen earlier, are more consistent, produce better quality wines and are more disease resistant. The upsurge in interest in making bottle-fermented sparkling wines seems set to continue and Pinot noir must be considered as a major variety in this respect. Today's fungicides, especially for the control of botrytis are markedly better than they were only a decade ago and this undoubtedly allows growers to leave their grapes to hang for longer and to ripen more fully.

Whether or not global warming is affecting Britain now – or will do in years to come – is open to debate. However, most winegrowers sense that weather patterns are changing and we are now experiencing earlier bud-bursts (although this is coupled with a higher incidence of spring frost damage) and therefore longer growing seasons. This does mean that the more marginal varieties – of which group the Pinots and Chardonnay are members – may fare better in the next decade than the last. One other factor to be considered is that of marketing. There is no doubt that the public will buy wines bearing names that they recognise, and Pinot noir is one that has its devotees.

The upsurge in plantings since 1984, when 11-ha were recorded as being planted in vineyards in Britain, via 46.6-ha in 2002, to today's total of 1,370-ha and occupying 29.45 per cent of the planted area, is impressive and speaks volumes about the ease with which it can be grown, its relatively good crops and its suitability for turning into fine sparkling wines. It is also due to three other factors: what I have called the 'Nyetimber effect', the weather and the name. The first factor showed that good, even great, sparkling wines could be made using the classic varieties, of which Pinot noir was an important member (although the first Nyetimber release was actually a *blanc de blancs* made from 100 per cent Chardonnay). Since then, all of the major players in Britain's sparkling wine business have shown that this was no fluke. The second factor, the weather, has meant that the acids are manageable and – if you want some colour for making rosé wines – the colours are acceptable. It has also meant that it crops adequately – another not-to-be-forgotten aspect of the variety. Thirdly, the very name 'Pinot noir' has a magic which cannot be denied, partly because it can make some of the best wines in the world, but also because it is such a chameleon, making both great red still wines and great (white) sparkling wines. Everyone talks about it being a 'difficult' variety and a 'heartbreaker' (whatever these descriptions actually mean) which further adds to its mystique. It is also a name which almost every wine drinker will know.

On the red wine front, I am more convinced now than I used to be about its suitability to make good red wines in Britain and in the last ten years there have been some superb examples from the top producers. As I said earlier in this book, in the national 'Wine of the Year Competition' wines made from Pinot noir have won the 'Bernard Theobald Trophy for the Best Red' a total of 23 times out of the 33 times it has been awarded, with Gusbourne (6 times) and Chapel Down (5 times) being the most consistent winners. Since the competition started in 1973, wines made from Pinot noir have won either gold or silver medals a total of 56 times, a total far, far higher than any other red variety. The top Pinot noir producers show that it is possible to make serious wines from this variety in most vintages and to make wines that sell for serious prices: £35–£45 a bottle.

The best Pinot noir wines are generally quite light in colour (although no lighter than many top-class Burgundies) and have simple red-fruit bouquets – perhaps with a touch of oak – adequate tannins and a fresh, fruity crispness about them that says 'attractive' rather than 'serious'. They are usually quite light wines and must be put in context with wines from other cooler growing regions: the Loire (Sancerre especially), Burgundy and even Alsace. Against wines from warmer climes they do not show quite so well, although the odds are shorter now than they used to be. Things (as they say) can only get better.

However, despite these successes, too many of the Pinot noir wines made in GB are thin and acidic and poor value for money. Many British-grown Pinot noirs would be much better presented as light, fruity rosés, maybe with a gentle touch of sweetness and a very light *petillance*, than as half-hearted, astringent reds.

Viticulturally, Pinot noir presents no great problems to the seasoned grower. It buds up quite early, making it susceptible to damage on frost-prone sites, but it will shoot from secondary buds. It gets all the usual ailments and requires good management to keep it *botrytis*-free, an important factor if the grapes are needed for whole-bunch pressing or for fermenting on the skins as a red wine. With a good spray regime and deleafing at the appropriate times, 100 per cent clean fruit is achievable in almost every year. On the cropping front, I have been very impressed by the level of the yields and while cooler years with difficult flowering conditions might set the averages back, 10 tonnes-ha (4 tonnes-acre) would appear to be achievable in 'normal' years and even higher in better years, which is fine for sparkling wine. For still red wine production, the quality of the site and the year will play a major rôle in fruit quality, but it is probably wiser to aim for nearer half this figure if good wines are to be made.

Clonally, Pinot noir is very diverse and I have no doubt that in Britain we have only scratched the surface of this aspect of growing. It is a great pity that very little work (i.e. none) has been done in Britain on clones suitable for our unique growing conditions (another job for Plumpton College perhaps?). We can only, therefore, take anecdotal evidence gleaned from existing growers in Britain, and take notice of recommendations offered by overseas growers.

Pinot noir is undoubtedly here to stay and, given its success in making some great sparkling wines and an ever-increasing number of really good still red wines, I can only see its importance to growers in Britain increasing

Note: See the separate section at the end of this chapter on: *Clones of Chardonnay, Pinot noir and Meunier*

Pinot noir Précoce

For the above variety see entry for Frühburgunder, Blauer

Reichensteiner

Type: *Vinifera*
Colour: White
Origin: Geisenheim, Germany
Crossing: Müller-Thurgau x (Madeleine Angevine x Calabreser-Froelich)
Date of crossing: 1939
Year of registration: 1978
Planted area in Germany (2024): 30-ha
Planted area in GB (2024): 64.53-ha

Reichensteiner is a Geisenheim cross produced by Professor Henrich Birk in 1939 by crossing Müller-Thurgau with a cross of two other varieties: Madeleine Angevine, the female French table grape used by Scheu (among others) for a number of his crossings, and Calabreser-Froelich, an early white Italian table grape. Together, these have combined to produce an early ripening variety, capable of producing large crops of relatively neutral grapes, high in natural sugars and low(ish) in acidity.

In Germany, Reichensteiner was seen as a substitute to the practice of 'wet-sugaring' (where 10 or 20 per cent by volume of water could be added to the must during the enrichment process in order to lower the acidity) – a practice that only came to an end in the late 1970s. In Germany, the area of Reichensteiner has now fallen to only 30.0-ha in 2024 and it is certainly not being planted there anymore. New Zealand once had 74-ha of it (where it regularly cropped at over 20 tonnes-ha) but it is now no longer listed as a separate variety in their annual vintage reports. It is however found in many of the cooler vinegrowing regions – Belgium, Canada, Denmark, the Netherlands and Sweden – where its relatively high yields (higher than more classic varieties) mean it is an economically viable variety. Its low acidity means it can be used as a blending partner for varieties with more acidity. It has also been used as one of the parents in several crossings with Gamay such as Garanoir and Ehrenbreitsteiner.

Reichensteiner was introduced into Britain by Jack Ward when he extended his Brickyard vineyard in Horam in 1971. Early results showed that it was a consistent cropper, with good yields and high sugars. Although it can be vigorous in its youth, it settles down after a few years and is less vigorous than MT at the same age. It has a good open habit and a leaf-wall that is less crowded and better ventilated than MT, both of which help disease control.

As a varietal wine, it does not have the character of fruitier varieties, but has lower acid levels, more body and extract, and higher natural sugars and is probably best used in a blend. If picked early, when acidities are higher, it can be used for sparkling wines with good results but, if left to fully ripen, it can also be used for the production of *süss-reserve*. Although it has fallen in area, from 113.9-ha in 1990 to around 64.53-ha in 2024, it has now started to expand its plantings and I believe it still has a valuable part to play in Britain's wine industry on account of its reliable cropping record. It is certainly a variety that the more climatically challenged regions should not ignore. Vines are difficult to obtain and you may find you need to have them custom-grafted i.e. produced to order. I also wish the name was a bit less Teutonic! Given that GB is the largest grower of this variety, maybe we could apply for a change in name?

Sauvignon blanc

Type: *Vinifera*
Colour: White
Origin: Original variety
Planted area in Germany (2024): 2,054-ha
Planted area in GB (2024): 56.17-ha

Sauvignon blanc is one of the world's most popular and widespread white vine varieties and can be found in climates both cool and warm. Great wines can be produced in a variety of styles: Marlborough's high-acid, zesty pure-fruit wines; the classic lean and minimalist wines from Sancerre and Pouilly-Fumé; the full-bodied, oaky offerings labelled as Fumé Blanc from California; and of course, a major component of most sweet wines from Sauternes and Barsac and other appellations in south-west France. These are all fine examples of this versatile variety. In Germany it has risen in popularity from nothing in 1999 to 2,054-ha in 2024 and the area being planted is still rising. It is also popular in Austria, northern Italy and the Adriatic countries and is just starting to be grown in Luxembourg. There are numerous Sauvignon blanc clones – 21 listed for use in France alone – and plenty more listed for use in Italy. Ninety per cent of New Zealand's Sauvignon blanc is the Californian UCD1 clone (said to have originated at Château d'Yquem in the 1880s) and Riversun, the major NZ vine nursery, lists a further seven clones as being of interest.

In Britain, Sauvignon blanc has risen in area from 0.5-ha in 1990, via 4.6-ha in 2013, 17.34-ha in 2022 and now (2024) stands at 56.17-ha. This relatively large (in percentage terms) and rapid rise has been fuelled by some substantial plantings in Essex, Kent and East Sussex. In these prime regions I guess it has a chance to shine, more so than the 5.30-ha planted in Cornwall where it will not like the wet and the wind. My experience with it has been slightly negative and, except on the very best sites and in the very warmest of vintages, it struggles to set fruit and struggles to ripen that fruit. Denbies, who claimed, in 2010 when they planted 1.21-ha of it, that it was a 'British first' (it was not), were very bullish about their 2013 harvest, which was picked on 31 October at 9.2 per cent, 11.4 g/l acidity and had a pH of 2.94. These are exceptionally good figures (especially given the year). The wine, however, was very grassy and perhaps 2013 was not the year for it. They ended up blending it with Bacchus. In the end they grubbed the vines up and as their winemaker, John Worontschak said, 'Wrong clone, wrong rootstock, wrong country'. In more recent times, Woodchester Valley Vineyard, near Stroud in Gloucestershire, won an IWSC Gold and a Gold at the Drinks Business 'Global Masters' competition for their 2018 Sauvignon blanc, which shows that it can be successful, but which rather goes to confirm my point about it needing an exceptional vintage for it to ripen. Ideally, if Sauvignon blanc is to be successful in Britain, someone needs to do some clonal trials as, out of the many dozens of different clones grown around the world, there are bound to be some more suitable for our conditions than others.

As a commercial variety for Britain I am still very doubtful about Sauvignon blanc. But of course it has a recognised name and style, always an advantage when it comes to selling wine. Maybe blended with, say, Bacchus in poorer vintages or even with another, more neutral variety, it will find a niche. I used to say that Chardonnay was impossible in Britain for still wine, so who knows? Maybe in fifty years' time, Britain will be growing all the Sauvignon blanc and New Zealand will be sending us table grapes!

Schönburger

Type: *Vinifera*
Colour: White
Origin: Geisenheim
Crossing: Pinot noir x Piròvano 1
Date of crossing: 1939
Year of registration: 1980
Planted area in Germany (2024): 13-ha
Planted area in GB (2024): 19.92-ha

Schönburger is a crossing made at Geisenheim by Professor Dr Heinrich Birk in 1939 and first registered for use in 1980. Like many of the crosses from this era, it combined a classic winemaking variety, Pinot noir, with an early ripening table grape variety, in this case a crossing from the Italian vine breeder Alberto Piròvano (1884 -1973) called Piròvano 1 which is Chasselas rosa x muscat Hamburg. The hope was that a unique German table grape could be produced. There is very little Schönburger planted in Germany, only 13.00-ha in 2022, down from previous plantings, and countries such as Canada, New Zealand and Britain have more. Some can even be found in Australia and South Africa!

It has been grown in Britain since the late 1970s and on account of its high-quality grapes, its ability to be turned into award-winning wines, its low acidity and its regular crops, it became quite widely planted. Colin Gillespie at Wootton Vineyard made it his speciality and won the Gore-Browne Trophy with it in 1982, and Lamberhurst won the same trophy with Schönburger wines in 1982 and 1990. By 1990 there were 75.3-ha, making it then the fifth most planted variety. It is an undemanding variety, more disease-resistant than MT and not so on-off in its cropping habits. The grapes are often very small. Yields are never large, but what is picked is almost always of excellent quality, with high sugars and low acids. As they ripen, the grapes change colour from a light lime-green to, at first, light pink and then, when fully ripe, to an almost tawny brown colour. This gives a very good visual indication to the pickers as to which bunches are ripe and allows for some selective picking. This can only aid wine quality. The wines are almost always light and very fruity, with some good muscat tones (some resemble a less powerful version of Gewürztraminer) that are best balanced with a slight amount of residual sugar. These wines age well and have been known to keep for up to seven years. Schönburger is one of the few varieties we grow that can be eaten with pleasure and although the berries are small and have seeds, they are packed with flavour.

Despite all these attributes, the area being grown in GB has fallen to just under 20.0-ha. Why? Well the name is tricky – an umlaut and '—burger' on the end never helped any vine variety become popular – and in style it is just a bit too – well, it has to be said – a bit too weak and wishy-washy[35]. I say this even though a well-made Schönburger, with a touch of residual sweetness, nicely chilled and served as an apéritif, is as good an English wine experience as you could want – not serious, but light, and slightly frivolous. Why has this counted against it? Who knows? Fashion in wine is a fickle thing and while I like Gewürztraminer, Viognier, Torrontes, and other aromatic, flowery varieties, I know that I am in the minority. However, I do sense that there is an increasing interest by producers in still wines, mainly because the cash-flow is so much better than long-aged sparkling wines. Maybe Schönburger is due for a modest revival? It's definitely on my 'ones to watch' list.

Siegerrebe

Type: *Vinifera*
Colour: White
Origin: Alzey, Germany
Crossing: Freely pollinated seedling of Madeleine Angevine
Date of crossing: 1929
Year of registration: 1958
Planted area in Germany (2024): 55-ha
Planted area in GB (2024): 18.34-ha

Siegerrebe is another Georg Scheu crossing from Alzey, made in 1929 and released to growers in 1958. In the *Taschenbuch der Rebsorten* (2010 edition) the breeding details are given as follows. It was originally said to be Madeleine Angevine x Gewürztraminer, but it was later revealed by Scheu's son, Heinz, to be a freely pollinated seedling of Madeleine Angevine (which is the same as Alzey 7672). More recently it has been confirmed in VIVC as Madeleine Angevine x Gewürztraminer. However, in *Wine Grapes* it is given as Madeleine Angevine x Savagnin Rose. Savagnin Rose (without an é) is a 'non-aromatic version of Gewürztraminer', which, in Germany, is often called Roter Traminer. In Germany there were 55-ha of Siegerrebe in 2024, down from 75.0-ha in 2018. Siegerrebe can reach very high natural sugar levels and in Germany is often used for making sweet dessert wines, even though the acid levels can be very low. The grapes have a strong muscat character and when ripe, can be very concentrated and almost overpowering. It is more often used for blending with other, less distinctive varieties.

It has been grown in Britain since 1957 when Brock was sent cuttings from Alzey. By 1960 he was able to report 'exceptionally early and the grapes have a strong bouquet' and he highly recommended it as suitable for both the table and winemaking. Despite its attributes of early ripening, low acidity and strongly flavoured grapes, it has really failed to be planted in

35 In previous editions I said that the variety was a bit too 'girlie' but bowing to current norms, I apologise for my lack of PC-ness and have changed the wording. The sentiment remains however.

anything like serious amounts and although one or two vineyards do make interesting varietal wines from it, it mainly gets lost in blends. Maybe because it ripens very early and falls so far outside what most growers consider their normal harvest time, it is seen as rather disruptive. Wasps are a particular problem and good wasp nest control needs to be practised if real damage is to be avoided. Birds, likewise, tend to be a nuisance. Three Choirs consistently make an interesting semi-sweet wine with the variety (and it is usually one of their better wines). Their 2006 Estate Reserve Siegerrebe won the Gore-Browne Trophy in 2008, their 2007 Siegerrebe won a gold medal in 2009 and their 2010, 2011 and 2012 all won silver medals. The area planted in Britain rose to 19.65-ha by 2015, but since then it has fallen back, albeit slowly, to 18.34-ha in 2024. This makes me think that it does not have much of a future. It is a pity as Siegerrebe can add a nice piquancy to an otherwise dull blend and is worth considering as a minor variety, especially if direct sales are your thing. As with several other German varieties, the name needs anglicising.

Chapter 6
Clones of Chardonnay, Pinot noir and Meunier

Until Nyetimber started winning medals and prizes (in 1997), the question of which clones of the Champagne varieties to plant had never really been discussed much among vineyard owners in Britain. In around 1998, when I was winemaking at Chapel Down, I planted 1.77-ha (4.37-acres) of Pinot noir and selected clones 115, 666, 667 and the *teinturier* clone Tête du Nègre.[36] The intention was to produce red wine, not sparkling, and this was the first time I had really given much thought to clones. The next occasion I had to choose some clones was in 2001–2 when I was asked to advise on the planting of the initial vineyards at Hush Heath Estate (now called Balfour Winery) which were to be solely for the production of a bottle-fermented rosé sparkling wine which came to be called *Balfour Brut Rosé.* This time I made sure I had done some homework before I put pen to paper. I therefore did as much investigative work as I could with: (a) the vineyards in Britain then making the best sparkling wine, (b) the CIVC, and (c) some of the *pépiniéristes* selling vines to Champagne and other sparkling wine producers.

From sparkling wine producers in Britain, principally Chapel Down, Nyetimber, and Ridgeview, I discovered some of the clones they were using, although about others there was a certain amount of secrecy. The CIVC sold me their *Viticulture raisonée guide pratique* which, in a separate chapter headed 'Clones', lists all the clones allowed to be used in Champagne: eleven clones of Chardonnay, eighteen clones of Pinot noir, and eleven of Meunier, all listed according to their ability (or otherwise) to crop, produce sugar and succumb to *botrytis*. From vine nurseries I was able to find out a bit more about other clones, especially ones with lower acid levels and better disease resistance, two attributes I felt were important in Britain. As with many other things viticultural, one always has to remember that when taking data from abroad, Britain is different, in some respects very different, from other vinegrowing regions.

Compared to Champagne for instance, yields in Britain are very much smaller. The average yield in Champagne since 2000 has been near to 15.00 t-ha (6.00 tonnes-acre) whereas in Britain, in the vineyards surveyed between 2016 and 2024 it has been 6.11 t-ha (2.47 tonnes-acre) for the Champagne varieties. Of course, the planting density in Champagne of around 8,000 vines-ha (3,238 vines-acre) compared to a typical British vineyard of around 3,500–4,000 vines-ha (1,416–1,619 vines-acre) makes a huge difference and the poor siting and amateur nature of many of Britain's smaller vineyards accounts in part for our low average yields. However, the truth is that high yields are what the Champenoise want and high yields is what they get. Their selection of clones therefore takes this into account.

In Britain, the natural sugar levels we achieve are, in all but the best of years, towards the lower end of those achieved in other cool climate regions and distinctly off the scale compared to anywhere warmer. Chardonnay and Pinot noir is generally picked at between 8.5 and 10 per cent natural alcohol, although in recent warm years – 2003, 2014, 2018 and 2023 – sugar levels have been at least 10 per cent and, for still wines, even higher. However, these are still low levels and make many an overseas grower wonder how we manage to make wine so good! Likewise, our acidity levels. Even in Germany where Riesling's high acids can test even the best winemakers, Britain's acid levels which, with Chardonnay, can often be around 14–16 g/l as tartaric, with pH levels routinely below 3.0, take a bit of getting used to. In Champagne, natural alcohol levels are around 10.00 to 10.50

36 *Teinturier*, meaning 'to dye or stain' in French, are clones that have deep red skins, plus red flesh of varying degrees of colour. They are usually a 'sport' or mutation of a standard variety, in this case, Pinot noir.

per cent,[37] with acidity levels (as tartaric) 11.00 to 11.50 grammes per litre, figures that are not too dissimilar to those achieved on recent years in vineyards in Britain, although, of course, one has to remember that yields in Champagne are often over three times ours.

One also has to remember when looking at clones of two of the three varieties in question – Chardonnay and Pinot noir – that these are also major varieties for the production of still wines of different qualities: high-volume, entry-level quality; mid-range, generally higher volume quality; and low-volume, top-notch quality. Chardonnay and Pinot noir are also varieties found in some of the warmest parts of the winegrowing world – California's Central Valley and some of Australia and South Africa's hottest regions – as well as some of its coolest, Champagne, Chablis and of course, coolest of all, Britain. Therefore, opinions about which clones suit which regions are always coloured by the type, style, and quality of wine being produced, plus the yield levels wished for.

Owing to our historic links with Germany in relation to viticulture, some vine suppliers seem keen to recommend and supply German clones grafted on to rootstocks common to German vine nurseries. This practice is something that concerns me. I take the view that if you are trying to produce a top-quality sparkling wine, then you have to have a very good reason not to copy what happens in the region that produces, without any argument, the best bottle-fermented sparkling wines in the world: Champagne. Champagne has the advantages of scale – there are around 35,000-ha (86,500-acres) of vines in the region – and of time – they have been growing vines in the region for many centuries. They also have the unique advantage that every individual vinegrower, all 15,000 of them, growing vines in 270,000 individual plots, belong to one organisation – the CIVC – whose job it is to oversee every aspect of production, including research into what clones suit the wine being produced. Therefore, my practice since 2001-02 has been to stick very firmly to selecting French clones – mainly, but not only, CIVC-recommended clones – for producing sparkling wines in Britain.

Although I am sure there are very many good German clones of both Chardonnay and Pinot noir, there are several factors which make me wary of using them in Britain. To start with, Chardonnay is still a relatively new variety in Germany, and although they had 3,050-ha (7,537-acres) in 2024, making it the sixth most widely planted white variety (up from fifteenth place in 2013), most of it is used for making still wines and very little is used for sparkling wines. Likewise, Pinot noir. Although a more popular variety than Chardonnay, the 11,519-ha (28,463-acres) grown in 2024 make it the most widely grown red variety and second most widely grown variety overall. However, very little of it goes into bottle-fermented sparkling wines.[38] Although there are some very serious growers of Pinot noir, mainly in the southern growing areas, Baden especially, producing some excellent Burgundy-style red wines, much of the Pinot noir is used for making relatively light-character, fruity wine, which finds favour with the German consumer. An additional factor in Germany is the cost of production, and yields, both of which are high. For a vine variety and/or clone to be taken up in Germany, it needs to be easy to grow, show good disease resistance and yield well. Low-yielding, disease-prone clones are soon weeded out.

Clonal development of Pinot noir in Germany started in the 1930s in three principal institutes: Freiburg, Geisenheim and Weinsburg. To start with, high yields and high natural sugars were the key goals, with disease resistance a distant

37 It is, however, worth noting that until the mid-'90s, grapes in Champagne only had to reach a potential alcohol level of 7 per cent before they could be harvested and that today it is only 9 or 9.5 per cent, depending on the vintage.

38 I visited Volker Raumland, who produces what are probably the best Champagne-style sparkling wines in Germany and asked him which clones he uses. He looked at me a bit quizzically and said he 'wasn't sure, but they were all French'.

aspiration. In the second generation of German clones, those with loose bunches were preferred as they had a greater resistance to *botrytis* and these have found favour with growers. Consequently, most German vine nurseries concentrate on producing these clones, and Freiburg and Geisenheim clones such as Fr 18-01 and Gm 20-13 are the most widely planted. Taking all these factors together, I am not persuaded that German clones are to be preferred over French ones for the production of sparkling wines in Britain.

As for recommending which clones to plant, as I have already said above, this is a confusing and unfathomable task. The clones I recommend in this section are ones used by many of the most important still and sparkling wine producers in Britain and all can be planted with confidence on suitable sites. This is not to say that they are the only clones that can be successful, and I am sure, in time, others will be tried and will prove suitable. My advice would be to plant at least two clones of each of varieties you have chosen, and on a vineyard of more than 2-ha (5 acres), you might plant more clones. On a small vineyard, around 0.4-ha (1-acre) of each clone makes a good-sized block and, if you are making your own wine, allows you to make separate *cuvées* of each clone. Of course, on a much larger vineyard, your blocks and areas of each clone will be much larger.

As a general rule, clones will, in most instances, vary only a little. Some will be earlier than others, some will have bigger and/or better crops than others, but these differences might well be much less important than the influences of the site, the rootstock, the weather during the year and of the management of the vineyard in all its aspects. Finding THE clone for your vineyard, one that will produce you perfect wine, year in and year out, is probably not possible, so play safe, and use a sensible range for the size of vineyard. Over the years, on your site, you will get to notice the differences and when you come to plant more, you can take a more educated decision. For still wines from Chardonnay and Pinot noir, much of the quality of the grapes will depend on the quality of the site, the yield level, canopy management and the degree of risk you are prepared to accept to leave the grapes to ripen

Chardonnay – still wine

For still wines, I regard clones 75, 76, 95 and 548 as suitable for Britain. Clones 121 and 131 have also been recommended for still wine, but acids are said to be higher. However, given that, in France alone, thirty-one clones of Chardonnay are registered for use and in other countries there are tens of dozens more (some of them duplicates, admittedly), I am sure there are other clones suitable for our vineyards yet to be trialled and judged to be suitable.

Chardonnay – sparkling wine

The following CIVC clones have performed well in vineyards in Britain for sparkling wine: 75, 76, 78, 95, 96, 118, 121, 124, and 131. I also like these non-CIVC clones: 119, 277 and 548. Clone 809, which has a slight muscat tone, might also add some interest to a blend and is also used by Gusbourne to make a well regarded still wine.

Meunier – sparkling and still wine

The following CIVC clones have performed well in vineyards in Britain for sparkling wine: 817, 865, 900, 977 and 978. Very little still wine has been made from Meunier, and where it has been, it has been made from sparkling wine clones. As with all still red wines, crop level, canopy management and harvest date are the keys to success.

Pinot noir clones in general

Pinot noir clones are loosely divided into four main types:

- Clones with compact bunches, also often with small berries. These are favoured by still red wine producers. Yields tend to be lower than other clones. They are also called *Pinot Fin* clones. 375, 667 and 777 are the main ones planted in GB vineyards.

- Clones with loose bunches, often with larger berries, and higher yields. German growers have planted these widely as they are more *botrytis* resistant and probably not the best for sparkling wines. They are also sometimes known as Mariafeld or Wädenswil clones which stem from Switzerland. Pépiniéristes Guillaume have their own proprietary clone of loose-bunched Pinot noir called their GL (*Grappe lâche*) clone.
- Upright clones (*Pinot droit*) have more erect foliage, favoured by some growers, and often have tight bunches. They tend to be less *botrytis*-prone. Good for sparkling wines, although can be later to ripen than other clones.
- Clones with small berries. German research stations (principally Geisenheim and Freiburg) have developed these clones which tend to have large bunches and high yields. They are good for still red wine in the German style, but are not much favoured in other regions.
- There are also several 'mixed' clones, as well as various *selection massale*[39] which many Burgundian growers now prefer.

The question of *botrytis* resistance has to be understood and taken into account when considering descriptions and recommendations in publications and on websites, all of which are based upon non-British conditions. As I point out in *Chapter 16: Pest and disease control* of my book *Wine Growing in Great Britain*, the incidence of *botrytis* in well-managed British vineyards is actually far less than it is in the regions of Germany and France where they grow Pinot noir. In Germany, for instance, where yields are high and grapes are often very tightly packed together inside the canopy, *botrytis* is a major concern. Some German growers practise *traubenteilen* – the cutting of the bunch in half at *véraison* – which causes the shoulders of the bunch to droop, loosening the bunch and opening it up. Many clones have therefore been selected for their resistance to *botrytis*, mainly based upon the clone having large, loose bunches, so that individual grapes are not packed tightly together, and air and light, plus chemical sprays, can penetrate the bunch easily. In Britain, I believe there are several reasons why *botrytis* is less of a problem than in warmer regions. To start with, it is generally windier in our vineyards which helps with drying the grapes out. Sugars develop later in the season when the night-time temperatures are lower, thus slowing down the growth of *botrytis*. Our acid levels are higher and pH levels lower as the grapes ripen, which again, slows down *botrytis* growth. Taken together, these all help to keep bunches free of *botrytis*.

Another aspect which also needs to be borne in mind when looking at Pinot noir clones is the question of colour. Deep-coloured red Pinot noir wine is not something that traditional Burgundian growers set much store by and in fact many old hands in the wine trade, on seeing a dark red Burgundy, will immediately suspect that it has been 'Hermitaged' – the practice of beefing up wines by the addition of Syrah-based Hermitage wine. Many German and New World growers believe that light coloured, slightly brick-red Pinot noir wines are to be avoided and thus select clones giving good colour. They also like clones producing wines with plenty of good fruity characters and one often sees the word 'cherry' mentioned in relation to non-French clones. Whether British growers require *botrytis* resistant Pinot noir clones producing fruity, dark red wine is open to question. For the production of high-acid, white sparkling wine, the answer is

39 *Selection massale* vines are produced by taking scion wood from (usually old) vineyards which are not planted with clones and contain vines whose ancestry is typically from a local successful vineyard. In this way, clonal variation and local typicity are maintained in new vineyards.

probably 'no'. For non-French clones, the German nursery Reben Sibbus has some interesting data.

Pinot noir – still wine

For still red wines, the most successful Pinot noir clones in vineyards in Britain appear to be the compact-bunch French clones 113, 114, 115, 375, 667 and 777. There are also good reports of two Geisenheim clones, Gm 20-13 and Gm 20-19, and an Italian clone, SMA 201. I personally prefer 667 and 777. There is also a *teinturier* clone of Pinot noir, called (not unsurprisingly) Pinot teinturier – also known as *Pinot tête-de-nègre* – which a few British vineyards have tried and which gives both good wine and very good colour. Grafted vines of these clones are hard to come by, but it is worth trying if still red wines are the aim. There are several sub-clones of *teinturier* clones including Bury, Guicherd, Mourot and Marsellein clones. The only one of these that I know is being used in GB is the Mourot clone.

If you start looking into Pinot noir clones, you will find a huge amount of information, much of it confusing and contradictory. Given that there are forty-five Pinot noir clones registered for use in France (including eighteen in Champagne), eight-hundred in total in collections in Alsace, Bourgogne and Champagne, dozens (if not hundreds) in other countries and that many nurseries have their own 'selections', this is a vast, unfathomable subject. Find a great site, get the density right, plant a couple of the clones recommended above, and when they start cropping, manage the yield and canopy properly and, given a fair year, you will produce something interesting – maybe a light red, perhaps a decent rosé. Only time will tell you if you have the right combination for great still red Pinot noir. Of course, with the 85:15 rule applying to labelling i.e. if the wine is at least 85 per cent of one variety, it can carry the name of that variety, there is ample scope to add a wine made from one of the better colour red varieties in order to beef up the colour of a Pinot noir varietal.

Pinot noir – sparkling wine

For sparkling wine, I am not as concerned about the question of clones as I am with still wine. There is so much that happens in sparkling wine production between picking and drinking that can substantially alter the nature, flavour, style and quality of the wine (blending, *sur latte* ageing and *dosage* being the main ones) that the clone is of secondary importance. It would be nice to plant those clones with the lowest acidity and highest yield and in time these will hopefully make themselves known. The following CIVC clones have performed well in British vineyards for sparkling wine: 292, 375, 386, 521, 779, 792, 828, 870, 871, 872 and 927. I also like non-CIVC clones 114, 115, and 459. Taittinger also planted a lot of 777 at Domaine Evremond (which surprised me) claiming that it was often used in Champagne (although it is not a CIVC clone!).

Note: The best source of information on French varieties, clones and rootstocks is the Pl@ntGrape website. The USA National Grape Registry, based at the University of California, also has a good website for varieties, although clones have all been given FPS (Foundation Plant Services) numbers: http://ngr.ucdavis.edu.

Chapter 7

PIWIs – Fungus resistant vine varieties

Introduction

Each vine is a new crossing. Geilweilerhof Aug 2025

PIWI stands for *Pilzwiderstandsfähig* (meaning 'fungus resistant' in German) and is used as a shorthand to describe new crosses of vines, mostly made since 1945, having both *vinifera* and non-*vinifera* parentage. Earlier multi-species varieties developed by hand pollinating the flower of one variety with the pollen of another are now generally referred to as 'hybrids' or sometimes 'interspecific crosses'. This booklet is not the place to go into the very complicated science behind plant breeding and genetics but suffice it to say that in the last twenty years the development of new vine varieties has become much more complicated, though with much better results[40].

Many people say, 'Why do we need new varieties?' This is indeed a very pertinent question. The foremost reason for producing both hybrids and PIWIs today is the desire by producers to grow vines that have a natural disease resistance to the ailments that affect vines of the species *Vitis vinifera*. The EU has indicated that agriculture in general must reduce its reliance on chemicals, and vines, as is well known, use a disproportionate amount of pesticides when compared to other crops. It is often quoted that vines, while occupying 7 per cent of agricultural land, use 70 per cent of all plant treatment products.[41] At the moment, the repository of disease repelling genes is held in wild vines whose grapes (if they have any) are not suitable for winemaking as they taste terrible. The dilemma – that the closer you are to wild vines the better the disease resistance is but the worse the wine tastes and that the closer you are to *viniferas* the better the wine tastes, but the worse the disease resistance is – is one that so far, has not yet been beaten completely. However, with each new generation of crossings the situation improves and the latest PIWIs are way ahead (in most cases) of previous ones. The current head of Geilweilerhof Institute of Grapevine Breeding, Dr. Oliver Trapp, told me that we need to look at the progression of new PIWI varieties and wine quality as we might look at the progression of our mobile phones. Take the late 19th century hybrids

40 The best place to get a full explanation of vine breeding in English that I have found is in 'The New Viticulture' pages 94-144 (Student Edition) by Dr Jamie Goode.

41 The comparison of the amount of pesticides used in the growing say of wheat and vines looks a little different if you compare them on a per hectare value basis. They might both have a yield of 10 tonnes per hectare, but ten tonnes of grapes can be worth ten to twenty times as much as the same amount of wheat. Also, different pesticides have very different rates of use. Sulphur, a contact material, is typically applied at between 5 and 10 kg-ha, whereas some systemic products only use 200 ml-ha. What metric do you use to compare one with the other?

that Kuhlmann (and others) produced. These are the equivalent of the first mobiles, the briefcase models. Then came the brick-sized phones. They are equivalent to the early PIWIs such as Regent and Rondo. Then take the best-selling phone of its time, the Nokia 1100. That's like Muscaris and Cabaret noir. Then onto the 1st iPhone – that is like Divico or Ravel blanc. Today, the best new varieties – such as Calardis Soleil, the latest PIWI from Geilweilerhof – are streets (or perhaps that ought to be 'fields') ahead of what has come before. Only crossed in 2010, it is already on the market and available. It has three genes against Powdery Mildew, two against Downy and one against Black Rot. No other variety has as many. The variety also has *Amurensis* parentage, Muscat parentage, and for good measure, Seyval blanc in its family tree. It is the iPhone 17 of its day. (See Figure 3).

Some history

The selection of plants for certain qualities was almost certainly something the first humans knew about. Mr Caveman (or probably Mrs Cavewoman) would have quickly realised that a certain tree or bush produced bigger and better berries or nuts, that certain wild grasses and plants produced bigger and better grains and beans (and more of them) and started the process of plant selection. They would of course have been looking for much the same things as today's plant breeders are searching for: better yields, natural resistance to pests and diseases; and better suitability to the environment they were growing in. The development of vines, until the arrival of Powdery Mildew (Oidium[42]) in Europe in 1830 or 1831[43], was one of natural selection by gardeners, plantsmen and those working in the vineyard and grape-growing business. They noticed which plants had the biggest bunches, which were the earliest to ripen and which were the tastiest. And therefore, when it came to planting new vineyards, they made sure they selected wood from their best vines from which to take cuttings in order to propagate new vines to plant. My second oldest book on vinegrowing, the 1662 edition of *The English Vineyard Vindicated* by John Rose, Gardener to His Majesty at his Royal Garden in St James's, gives several recommendations as to which varieties to plant. Mr Rose says that 'upon long and diligent observation of the Species, I do recommend these following' and then lists six varieties – the black Cluster grape, the white Muscadine, the Parsely grape, the Muscadella, the Frontiniaq and 'a new white grape' which he found in 'His Majesties Garden'. Right at the back of the book, in handwriting on the back of the last page, someone has written out a list of seven varieties – the Black Cluster, Lady Cranbrook's small Muscat, Mr Rose's early grape etc – which I assume a previous owner of this book had written in there, maybe when discussing with another grower what their favourite varieties were. In this way the practice of relying on certain varieties grew up and unlike plants grown from seed – which means almost all arable crops, grasses, and vegetables – the decision as to what variety to plant has to be taken with a very long-term view of the suitability of the species to the region it is growing in and to the style and quality of the wine made from it.

As Britain (and other western countries) started to enjoy the fruits of the Industrial Revolution, having a well-stocked garden complete with a greenhouse to grow tender plants, was very much part of your status. To do this, owners of stately homes dispatched plant hunters to all four corners of the (then-known) globe to gather specimens of different plants both decorative and practical. Joseph Banks (1743-1820), one of the foremost plant hunters of his day, is said to have

42 Powdery Mildew started out by being called *Oidium tuckeri*, then *Uncinula necator* and now *Erysiphe necator* I propose to stick to Powdery Mildew which is the most commonly used name for it in English.

43 There is some doubt about this date. It is generally stated that Powdery Mildew was discovered in 1845, yet the 1883 edition of Archibald Barron's 'Vines and Vine Culture' states that the Kentish Gazette of 1830 or 1831 had news of the outbreak.

brought back over 30,000 plant specimens during his lifetime. In 1768 he accompanied Captain James Cook on his first voyage to the southern hemisphere together with eight naturalists whom he (Banks) financed, and returned with thousands of plants. Of course, together with novel plants, the naturalists also imported novel pests and novel diseases, and it was in this way that both Powdery Mildew (in 1830-1) and Downy Mildew[44] (in 1878) were introduced to Europe, almost certainly from North America. Decorative vines such as Virginia Creeper, with their attractive autumn colours, were very popular to cover unsightly buildings and walls. Being from North America, where both Powdery and Downy Mildew were endemic, these vines had, over the millennia, developed a natural defence mechanism against both these diseases, unlike the European vine, *Vitis vinifera*. Seeing that the Virginia creeper vines (and other species of vine) were resistant to these two very destructive diseases, vine breeders soon came to the idea that maybe by crossing species of vines that were naturally resistant to these diseases, yet whose grapes were not very good for winemaking or eating, with those that were not disease-proof, but did have grapes that were good for winemaking and or eating, a new strain of vine with might be created. Thus were vines known as hybrids created.

The hybridisation of plants was nothing new – the first, a cross between a carnation and a sweet william and known as Fairchild's Mule, was created in 1716-17 – and rose breeders excelled at producing new and exciting cultivars. In this way, the first hybrid vines were bred to counteract the ravages of both mildews, but the discovery of *Phylloxera* in Europe in 1863[45] introduced a wholly new element into the hybridist's work. Initially the pest (or was it a disease?) baffled the experts appointed to find a cure for it. No male, yet able to reproduce? Developed root galls on viniferas but not on American vines, produced leaf galls on American vines but not on European ones? A very curious beast. However, one thing the hybridists did notice was that some of their crosses, bred to counteract mildew, were also more resistant to the depredations of 'the destroyer' than others. This led to the realisation that the three biggest maladies of the day to affect *Vitis vinifera* – two mildews and a root-sucking aphid – might be able to be controlled through plant breeding and not the sometimes mad-cap remedies which were initially suggested[46].

Early hybrids and interspecific crosses

The great hybridisers of the day – Couderc, Foëx, Gallard, Ganzin, Kuhlmann, Millardet, Oberlin, Ravat, Richter, Seibel, Seyve, and Villard – all contributed to developing both wine and table varieties, as well as rootstocks. There were battles to be fought between those that supported the 'HPD' (*hybrides producteurs directs*) producers who wanted to produce *Phylloxera* resistant vines that could be propagated from cuttings without the need of grafting, and the *greffageurs* – those that advocated the grafting of scion wood onto *Phylloxera* resistant rootstocks. Both types had their supporters, and both types their detractors. The hybridists themselves were pulled between the obvious benefits of being able to produce new vines without grafting – which also pleased vineyard owners – and the obvious benefits (to themselves) of having the production of both rootstock wood and grafted vines from which to make a living. Ultimately it was the grafters that won out, aided by the authorities in charge of setting the rules for wine production.

44 Downy Mildew, called *Mildou* in France, is also known as *Peronospera*. Its botanical name is *Plasmopora viticola*.

45 It was in a greenhouse in Hammersmith in 1863 that *Phylloxera* was first discovered in Europe, but it aroused very little interest. Its discovery however in 1863-5 in various vineyards in southern France created much more alarm and by 1868 a committee of the great and good of the vine world had been formed to find a cure, for which a substantial prize was on offer..

46 Flooding the affected vineyards, injecting highly poisonous carbon disulphide into the soil and burying a live toad under each vine – all were suggested as cures for *Phylloxera* but ultimately rejected.

Wine was too important not to be controlled and of course anything to do with alcohol, its production and sale, needed to be monitored and controlled. Following the financial and societal turmoil that *Phylloxera* had caused and the inevitable over-production that followed the systematic replanting of most vineyards in Europe, the French introduced rules about the provenance of wine. In 1935 they created the *Appellations d'Origine Controlee* system and in 1938, the Côtes du Rhône became the first wine region to get an AOC. One of the main tenets of the system was that wine from hybrids could not be classified as Quality Wine, and whatever the provenance or quality, it could only ever be Table Wine. This 'viticultural racism' (as the great vine breeder Helmut Becker used to call it), still exists (in the EU) and although the dam walls have broken down to a certain extent with the advent of PIWIs, older hybrids are still not (yet) permitted to be classified as Quality Wines in either the EU or Great Britain.

The early hybrids that became popular were those that delivered both quality (up to a point) and quantity (the more the better – especially for distillation). Varieties such as Baco noir and Baco blanc, both varieties used for the production of Cognac and other spirits, and varieties such as Chambourcin, Chancellor, De Chaunac, Léon Millot, Marechal Foch, and Seyval blanc, and multiple crosses from the vine-breeders of the day, mentioned at the start of this section, all originated in the period between 1870 and 1940 when almost all vineyards were being replanted because of *Phylloxera*. Many of the early hybrids, such as Concord, Catawba, Delaware, Isabella and Niagara, used *Vitis labrusca* varieties as breeding partners which, while giving good resistance, also had the classic 'foxy' hybrid taste that people disliked. Some of these older hybrids, while less popular than they were, can still be found in many parts of the world, especially in regions where winter temperatures are very low or where humid summers bring massive disease pressure.

Brazil for instance, a very hot and humid country, has over 80,000-ha of vineyards, only 15 per cent of which are planted with *viniferas*. The rest are planted with *labrusca* hybrids. Isabella is the biggest variety grown with over 13,500-ha planted, with Couderc noir, Concord, Niagara and Jacquez (all old hybrids) in the top 10. New York State has around 8,500-ha (21,000-acres) of Concord, much of it used for making juice and jelly, and in Canada, Vidal is the best variety for much of their ice wine.

This is not the place to recount the full story of the hybridists and their search for the perfect blend of American and European vines, but their success can be gauged by reference to the 1968 *Cadastre viticole*. This, the national register of all French vineyards, listed a total of 1,294,880 hectares (3.20 million acres) of wine vines in all regions except Champagne (which had banned hybrids in 1935) of which 406,000-ha or 31 per cent, were HPDs. And of course, not only did the hybridists produce grape-bearing vines. They also produced all of the *Phylloxera* resistant rootstocks that proved to be the only lasting practical solution to that major problem. But by 2010, the area of HPDs in France was down to 6,000-ha and reducing year by year.

Modern interspecific crosses and disease-resistant hybrid crosses (PIWIs)

New interspecific crosses developed from the early work that the French (and other countries') hybridisers did in the period prior to the 2nd World War. As has been discussed above, these crosses (with the benefit of hindsight) were relatively crude crosses made without the benefit of the DNA analysis to which present-day vine breeders have access. The first attempt at producing an inter-species disease resistant crossing was in 1936 by Prof. Bernhard Husfeld (1900-1970), then at the Kaiser Wilhelm Institute in Müncheberg, 50 km to the east of Berlin. This was a crossing of Oberlin 595 (Oberlin noir – a *Riparia* x Gamay cross) and Riesling and named, with due deference to the government of the day (the Nationalist Socialist German Worker's Party), Siegfriedrebe. With its good resistance against Powdery Mildew and a loose bunch formation, it became something of a hit at the time. Husfeld also produced Aris (in 1937), another Oberlin x Riesling cross, and two Oberlin x Foster's White Seedling crosses, Castor and Pollux (both in 1944), none of which gained much traction and were not widely planted, although small patches of these varieties can still be found around the world. As the Second World War entered its final phases and the Russians advanced on Berlin, the Kaiser Wilhelm Institute plant breeding division moved westwards to Siebeldingen, near Landau on the French border, taking the best of their new crosses with them. Here Husfeld founded the Geilweilerhof Institute of Grapevine Breeding[47]. From that date, crosses followed crosses, not only at Geilweilerhof but also at the other major plant breeding institutes in Germany: Freiburg, Geisenheim, Oppenheim, and Weinsberg.

In 1970, Professor Gerhardt Alleweldt took over the Geilweilerhof Institute and the breeding of resistant varieties moved up a gear. Regent (which is Diana x Chambourcin[48]), crossed by Alleweldt in 1967 and after considerable clonal selection and trialling, received variety protection in 1994, was registered on the national list in 1995, the EU list in 1996, and by 1998 was classified as available for use in most of the German winegrowing regions. Perhaps more important is that in 1996, Geilweilerhof requested that the *Bundessortenamt* (Federal Plant Variety Office) test Regent to see if its hybrid origins could be detected by comparing it to the attributes of *vinifera* varieties, except for its natural resistance to diseases (which it was, of course, vastly superior). Rondo, Orion and Phoenix were also subjected to the same tests.

The *Bundessortenamt* decided that they couldn't see any difference. Leaf and berry shape were *vinifera*-like, as was wine quality. Following this report, the four varieties were declared to be 'honorary' *viniferas* and added to the lists of those varieties allowed to be labelled as Quality Wines[49]. Regent, with 1,525-ha planted in Germany in 2024, is by far the most successful PIWI variety, although this is down from 2,047-ha in 2012. There are now well over 100 PIWIs registered for use in Germany, and probably another 50 or more in other European countries – both EU and non-EU.

What do you do when you grow 513 different white varieties? Make a Cuvée Blanc!

47 Now the Julius Kühn-Institut (JKI) for Grapevine Breeding.

48 Diana is Silvaner x Müller-Thurgau and Chambourcin is a complex Seyve-Villard and Seibel cross made by Johannes Seyve in the 1930s.

49 It took our Ministry of Agriculture, then MAFF, now DEFRA, five more years to agree that this was permitted in GB.

Vine-breeding establishments

References are made in the variety descriptions to the following vine-breeding establishments:

- Alzey, Rheinpfalz, Germany
- Blattner, Valentin, Soyhières, Switzerland
- Freiburg, Staatliches Weinbauinstitut, Germany
- Geisenheim Institute for Grapevine Breeding (Gm), Rheingau, Germany
- HBLA Klosterneuburg, Austria
- Julius Kühn Institute (JKI) for Grapevine Breeding, Geilweilerhof, Germany
- Mendel University, Brno, Czech Republic
- Oppenheim, Rheinhessen, Germany
- Potapenko All-Russia Research Institute, Rostov-on-Don, Russia
- Vivai Cooperativi Rauscedo (VCR), Italy
- Weinsberg, Baden-Württemberg, Germany
- Würzburg, Bayern, Germany

Breeding goals

Plant breeders working in the world of vines for wine production, have the following goals for any new variety:

- Outstanding wine quality
- Resistance to:
  - Powdery and Downy Mildew
  - Black Rot
  - *botrytis* bunch rot
- Climate adaptation
  - Resilience against:
    - Drought
    - Sunburn
    - Winter frost
- Growth habit
  - Upright growth habit to aid viticultural operations
  - Bunch architecture – loose bunches
  - Balanced yields
  - Good wood ripening

Chapter 8

Selecting PIWI and hybrid varieties to plant in Great Britain

Before you select a PIWI to plant, consider the following questions:

- **Question 1: Most widely planted PIWIs:** Are the most widely planted PIWI varieties the best?
- **Question 2: Wine quality:** Do PIWIs produce as good a wine as *vinifera* varieties?
- **Question 3: Disease resistance:** Can PIWIs be successfully grown without using any pesticides?

Question 1: Are the most widely planted PIWI varieties the best?

Most widely planted PIWIs and hybrids in Europe - 2025

Variety	Ha	Variety	Ha
Regent	1,607	Calardis blanc	91
Souvignier gris	1,143	Donauriesling	82
Solaris	427	Rondo (GM 6494/5)	80
Cabernet blanc	394	Phoenix	73
Muscaris	382	Pinotin	65
Roesler	284	Cabernet Jura	61
Johanniter	277	Rathay	55
Floréal	247	Satin noir (Estimate)	49
Sauvignac	219	Prior	48
Cabernet Cortis	154	Monarch	41
Seyval blanc	134	Helios	35
Blütenmuskateller	124	Bronner	29
Soreli	117	Cabertin	28
Divico	113	Caberet noir	28
Vidoc	112	Pinot Nova	19

Table 22

Most widely planted PIWIs and hybrids in Europe 2025

Obtaining reliable information about which varieties are the most widely planted is quite difficult in some countries. Germany, Austria and Switzerland (as one might expect) publish their data, and Belgium also provides good varietal data. France, Italy and Spain appear to be coy about their plantings of PIWIs. Reliable data from smaller northern European countries such as Denmark, Estonia, Finland, the Netherlands, Norway, Poland, and Sweden is more difficult to obtain. I have however, worked with vineyards in both Denmark and Norway, so have a fair idea of the situation in Scandinavia.

As can be seen from Table 22, Regent, one of the earliest PIWIs to be developed and released (crossed in 1967; released in 1994) tops the list of European PIWI plantings, but it is worth bearing in mind that in 2012 the area of this variety stood at 2,047-ha. Souvignier gris on the other hand in 2022 only stood at 205-ha, whereas in three years it has grown over five-fold to 1,143-ha and is undoubtedly going to go further. It is worth noting which varieties are on an upward trajectory and which on a downward one.

Germany

The twenty-four vine varieties planted in Germany at 500-ha or more (see Table 23) are comprised of traditional varietals, some standard international varieties that have become more important (the fruits of global warming), some pre-1945 *vinifera* crosses, and two PIWIs – Regent, which is slowly reducing in planted area and Souvignier gris, a more recent one, which is increasing. Riesling and Silvaner, which once accounted for 50 per cent and more of all German vineyards, are still there with Riesling at 24 per cent and rising slowly, but Silvaner is gently fading away. Chardonnay, Pinot noir, Pinot gris and Sauvignon blanc are expanding and together now account for 24 per cent of plantings. Crosses such as Müller-Thurgau (crossing made in 1882), Dornfelder, Kerner, Bacchus and Scheurebe continue to find favour with consumers and account for 20 per cent of plantings.

Germany - all varieties, 500-ha and over - 2024

Variety	Hectares	Percentage
Riesling	24,233	23.55%
Pinot noir	11,437	11.12%
Müller-Thurgau	10,511	10.22%
Pinot gris (Ruländer)	8,404	8.17%
Dornfelder	6,507	6.32%
Pinot blanc	6,362	6.18%
Silvaner	4,228	4.11%
Chardonnay	3,050	2.96%
Portugieser	2,107	2.05%
Sauvignon blanc	2,054	2.00%
Limburger	1,888	1.84%
Kerner	1,874	1.82%
Trollinger	1,791	1.74%
Schwarzriesling (Meunier)	1,598	1.55%
Regent	1,525	1.48%
Bacchus	1,515	1.47%
Scheurebe	1,503	1.46%
Traminer (Gewürztraminer)	1,115	1.08%
Gutedel (Chasselas)	1,025	1.00%
Merlot	963	0.94%
Muskateller	605	0.59%
Souvignier gris	596	0.58%
St Laurent	553	0.54%
Cabernet Sauvignon	516	0.50%
All other varieties	6,926	6.73%
Total Hectares	**102,886**	**100.00%**

Source: Deutches Weininstitut 2005

Table 23

PIWIs in Germany - 2024

Variety	Hectares	Percentage
White varieties		
Souvignier Gris	596	16.21%
Cabernet Blanc	335	9.11%
Solaris	230	6.26%
Sauvignac (Estimate)	200	5.44%
Muscaris	166	4.51%
Johanniter	155	4.22%
Calardis blanc (Estimate)	90	2.45%
Phoenix	43	1.17%
Helios	33	0.90%
Saphira	11	0.30%
Hibernal	10	0.27%
Bronner	7	0.19%
Merzling	3	0.08%
Villaris	1	0.03%
Total whites	**1,880**	51.13%
Red varieties		
Regent	1,525	41.47%
Cabernet Cortis	80	2.18%
Pinotin	53	1.44%
Satin noir (Estimate)	47	1.28%
Prior	26	0.71%
Cabertin	24	0.65%
Monarch	17	0.46%
Rondo	11	0.30%
Cabernet Carbon	6	0.16%
Piroso	4	0.11%
Cabernet Carol	4	0.11%
Total Reds	1,797	48.87%
Total Ha all PIWIs	**3,677**	**100.00%**

Source: Deutches Weininstitut 2025

Table 24

PIWIs in Germany

As can be seen in Table 24, the most widely planted PIWI variety by far in Germany is Regent, the first PIWI to be released for commercial planting (in 1994). Back in those dark pre-global warming days, it was just what German growers wanted in a vine variety: early ripening, easy to manage with its natural disease protection, acceptable wine quality, and heavy yielding. Boxes ticked all round.

With first plantings in 1995, by 1998 there were 187-ha of Regent in Germany, and by 2008 there were 2,161-ha. However, this has now fallen to 1,525-ha as better varieties take its place. This is undoubtedly due in part to climate change which has made varieties such as Cabernet Sauvignon and Merlot now commercially worth planting. These are both better known to consumers and produce better quality wine. Regent's modest quality really does not make up for the attributes of the lower costs of growing and the high yields of this variety.

When you look at what's happened in German vineyards since 1995, some of the answers to Regent's decline are clear. Overall, the German vineyard area is down from 105,634-ha in 1995 to 102,886-ha in 2024, a fall of 2,748-ha. White varieties fell by 14,094-ha (mainly Müller-Thurgau, Silvaner, Bacchus, and Scheurebe), while some white varieties (the more traditional ones) increased: Riesling, Pinot gris, Pinot blanc, and Chardonnay. But red varieties are up by 12,418-ha, the major varieties planted being Pinot noir, Dornfelder, and Merlot with the less widely grown (in Germany) varieties St Laurent, Cabernet Sauvignon, Acolon, Domina, Cabernet Mitos, Cabernet Dorsa and Blauer Frühburgunder (Pinot noir Précoce) all increasing by smaller quantities. PIWIs, while expanding in Germany, still only account for 3.6 per cent of all vineyards in Germany. White PIWIs appear to be faring better than reds which is an interesting fact. Probably because the wine quality of the whites is higher, or maybe because white wine drinkers are less demanding – who ever heard of a red wine *Schorle* (spritzer)?

To really understand the development of PIWIs in Germany (and indeed in other countries which have been growing them for more than 30 years) you need to factor in several things: the marketing elements that make a variety popular and sustainable (in the sense that growers want to plant it again once those vines have had their day); is it a variety that can achieve high (even very high) quality and therefore high prices; and what are the yields like? Or is it a variety that can be grown cheaply, with high yields, acceptable quality (for the price) and therefore be a profitable addition to the business?

One can see from the details of the varietal changes over the last 28 years, almost the lifetime of a modern vineyard, that climate change has allowed the planting of several full-bodied red varieties. These are more suited to the climate of today, and more recognisable to German consumers as quality varieties for which they are prepared to pay higher prices than for older, less fashionable white varieties such as Müller-Thurgau and Silvaner. Interestingly, the only two PIWIs to make it onto the top 30 list are Regent and Souvignier gris.

Austria

PIWIs in Austria 2023-24	
Variety	**Hectares**
Roesler	283.78
Blütenmuskateller	123.72
Muscaris	96.88
Donauriesling	80.80
Souvignier gris	75.62
Donauveltliner	62.35
Rathay	54.57
Cabernet blanc	24.33
Cabernet Jura	17.84
Pinot Nova	17.60
Bronner	16.75
Solaris	11.53
Johanniter	9.35
Regent	7.92
Total Ha	**883.04**

Table 25

With almost 25 per cent of its 44,210-ha of vines farmed organically (compared to Germany's 9 per cent), Austria has taken to PIWIs in a big way. As can be seen in Table 25, the total area of PIWIs is 883-ha and made up mainly of varieties crossed at HBLA Klosterneuburg with a scattering of Freiburg and Blattner varieties. Most of the varieties are quite old (in PIWI terms) dating from 1970-1990. The newest are Muscaris and Souvignier gris. The variety with the largest planted area is the red varietal Roesler, a crossing of Zweigeltrebe crossed with a crossing of Seyve-Villard 18-402 and Blaufränkisch.

Switzerland

Switzerland - Major varieties 2023

Variety	Ha	Variety	Ha
Pinot noir	3,702	Garanoir	226
Chasselas (Gutedel)	3,502	Sauvignon Blanc	219
Merlot	1,221	Syrah	210
Gamay	1,051	Cornalin (Landroter)	163
Gamaret	429	Humagne Rouge	143
Müller-Thurgau	425	Diolinoir	128
Chardonnay	410	Pinot blanc	112
Sylvaner	321	Divico	95
Petite Arvine	261	Cabernet Franc	87
Pinot gris	236	**Sub-total**	**13,169**
Savagnin Blanc	228	277 other varieties	1,400
		Total hectares	**14,569**

Source: L'Année Viticole 2023, Office Fédéral de l'Agriculture

Table 26

In 2023, Switzerland had over 14,500-ha of vines planted (see Table 26) with an amazingly diverse number of vine varieties. Over 300 different ones are recorded as having cropped in 2023, which therefore probably does not include any non-cropping vines from new plantings made in the six years between 2020 and 2025. The list is dominated by Pinot noir and Chasselas (also known as Fendant or Gutedel, depending on whether you are in the French speaking or German speaking part of the country), with Merlot and Gamay at positions three and four. Together, these four account for 65 per cent of all plantings. The only PIWI to make it into the top twenty is Divico, the home-grown red variety, which sneaks in with 95-ha.

Switzerland - PIWI & hybrids cropping in 2023

Variety	Ha	Variety	Ha
Divico	94.93	Bianca	3.11
Souvignier Gris	45.34	VB Cal 1-36	3.10
Johanniter	41.43	Cabernet Cantor	2.80
Cabernet Jura	40.71	Vidal Blanc	2.29
Solaris	36.68	Acolon	2.01
Muscaris	30.37	Birstaler Muskat	1.79
Regent	28.61	GF 48-12	1.61
Maréchal Foch	14.63	Merlot Khorus	1.60
Seyval Blanc	11.52	Baco Noir	1.54
VB Cal 1-28	10.66	VB 91-26-26	1.49
Cabernet Blanc	9.63	Catawba	1.34
Léon Millot	8.74	Donauriesling	1.24
Sauvignon Soyhières	7.41	Pinotin	1.18
Prior	6.92	Plantet (Seibel 5455)	1.10
Cabernet Cortis	5.27	VB Cal 1-20	1.10
Cabaret Noir	4.66	Pinot Nova	1.05
Isabella	4.65	Sauvignon Nepis	1.04
Chambourcin	3.88	Bronner	1.02
Cabertin	3.61	64 other varieties	43.94
Cabernet Soyhières	3.39	**Total**	**487.40**

Source: L'Année Viticole 2023, Office Fédéral de l'Agriculture

Table 27

As for PIWIs in Switzerland, they probably have the greatest number of different crossings than anywhere in the world. The records I have, again for vineyards that cropped in 2023, so undoubtedly missing several years' worth of plantings, show well over 100 PIWIs plus a few older hybrids being grown (see Table 27). In total there were 487-ha of PIWIs and hybrids, which is 3.3 per cent of the Swiss vineyard area, and undoubtedly getting larger. As has been said, Divico heads the list followed by a selection of the most popular ones such as Souvignier gris, Johanniter, Cabernet Jura and Solaris. Surprisingly, the old hybrids Maréchal Foch (with 14.63-ha) and Seyval blanc (11.52-ha) make it into the top ten varieties which account for 72 per cent of all plantings of PIWIs. There are also some old hybrids such as Léon Millot, Isabella, Chambourcin, Vidal blanc, Baco noir, and Catawba which put in an appearance lower down the list. Of course Switzerland, not being in the EU, has its own rules on vine varieties as in the EU Isabella and Catawba are not allowed to be grown at all! The popularity of PIWIs in Switzerland is also undoubtedly due to the fact that Valentin Blattner, nurseryman and PIWI producer *extraordinaire* lives in Switzerland.[50] He has an astonishing 10 varieties among the top forty – 23 percent – grown in Switzerland! The other reason is that Switzerland also has at least three institutions producing new varieties, both PIWIs and *viniferas*.

Belgium

Belgium has a history of both Roman and mediaeval monastic viticulture, very similar to that of GB's, plus it also had a substantial table grape industry which started in the 1860s, became very large in the 1920s with almost 300-ha of glasshouses, but since then, with the advent of refrigeration for shipping, transport and storage and in the home, has declined. Unlike GB, whose efforts of establishing vineyards for wine production have been expanding since the first modern vineyard was planted in 1952, the Belgians were slow to get away from the starting blocks and even today their vineyards only amount to 891-ha. However, with around 130 different grape varieties and an impressive selection of 41 PIWIs covering 264-ha (which is 30 per cent of their planted area) they are considerably ahead of GB in this respect. They also have no Seyval blanc, and very little planted of the old hybrids. Their 'top ten' PIWIs which are what one might now call the 'standard ones', account for 82 per cent of their PIWI vines.

Vine varieties in Belgium 2023

PIWI varieties - Ha		Vinifera varieties - Ha	
Johanniter	69.50	Chardonnay	289.96
Souvignier gris	33.36	Pinot noir	111.43
Solaris	31.76	Auxerrois	36.38
Muscaris	23.40	Meunier	32.52
Sauvignac	12.99	Pinot blanc	20.64
Regent	12.93	Müller-Thurgau	13.37
Cabernet Cortis	12.29	Riesling	11.05
Cabaret noir	8.16	Dornfelder	10.50
Pinotin	7.23	Acolon	7.90
Rondo	4.65	Merlot	5.82
Merlot Khorus	4.60	Gamay	4.34
Bronner	4.51	Kerner	4.11
Cabernet Cantor	4.45	Sauvignon blanc	3.43
Divico	4.36	Zweigeltrebe	2.77
Voltis	3.64	Gewürztraminer	2.67
Cabernet blanc	3.44	Früburgunder	2.51
Ravel blanc	2.96	Chenin	1.95
Muscat bleu	2.75	Gamaret	1.44
Satin noir	2.33	Siegerrebe	1.24
Cabernet Jura	2.03	Sauvignon gris	1.17
Helios	1.55	33 other viniferas	61.58
Phoenix	1.54	**Total ha of viniferas**	**626.78**
Pinot Kors	1.53	PIWIs	264.22
Monarch	1.52	**Total ha in Belgium**	**891.00**
Soreli	1.12	Source: La Viticulture Belge	
16 other PIWIs	5.62	SPF Economie 2023	
Total PIWIs	**264.22**		

Table 28

50 www.domaineblattner.ch

PIWIs in France

PIWIs in France (2021)	
Variety	**Hectares**
Souvignier gris	388
Floréal	247
Artaban	153
Soreli	116
Vidoc	112
Muscaris	60
Cabernet Cortis	50
Voltis	6
Monarch	22
Cabernet blanc	20
Prior	15
Total Ha	**1,189**

Table 29

France has had a resistant vine breeding programme running for over 50 years and this is now known as the INRA-ResDur programme. It released four new varieties in 2018 – Artaban, Floréal, Vidoc and Voltis – and five in 2021 – Coliris, Lilaro, Opalor, Selenor and Sirano. As you can see from Table 29 some of them have been taken up, but in a pretty small way. The French seem far less keen on them than their European cousins, but I suspect that it is the rigid *appellation* system that keeps them out. These new ResDur varieties seem mostly to be bred for hot growing conditions and most of the plantings are in the south and south-west, where organic and biodynamic growers are having problems controlling Downy Mildew with their annual allowance of copper. The new crosses are also being used to counter climate change with varieties that accumulate sugar more slowly so that alcohol levels are lower. In regions where natural alcohols in some varieties are now regularly getting into the 17-18 per cent ABV[51] region, selecting varieties with lower alcohol production is an obvious advantage. The only variety I can see which might be of interest to GB growers is Voltis, a variety approved for use in Champagne (under certain restrictions) and this has been included in my PIWI selection.

51 Alcohol by volume

Netherlands - 2023	
Major varieties	**Minor varieties**
Auxerrois	Cabaret noir
Cabernet Cortis	Cabernet blanc
Johanniter	Cabernet Cantor
Pinot noir	Calardis blanc
Regent	Merzling
Solaris	Monarch
Souvignier gris	Muscaris

Table 30

Netherlands

The Netherlands (or Holland as most Britons call it) has around 302-ha of vines spread around the hillier parts of the country, mainly in the south towards the Belgian and German borders. There are said to be 135 separate producers. In terms of vine varieties, their major plantings appear to be split between *viniferas* and PIWIs. Precise data seems to be unavailable (despite them being in the EU).

Minor European winegrowing countries

Vines are now grown in almost all European countries and in the cooler and more marginal ones – Finland, Denmark, Norway, and Sweden, plus Baltic Germany, Poland, and the Baltic states, Estonia, Latvia and Lithuania – PIWIs are important. In Scandinavia and the other cooler regions mentioned above, the PIWI varieties that appear to be the most successful, and by that, I mean produce a meaningful crop of grapes which stand a chance of ripening, are Rondo and

Solaris, with smaller areas of Cabernet blanc, Cabernet Cantor, Cabernet Cortis, Cabaret noir, Johanniter, Muscaris, Regent and Souvignier gris (in alphabetical order).

The Republic of Ireland has around 17 different vineyards spread all over the country with the oldest, the Thomas Walk Vineyard, established in mid-1980s planted on a sheltered, sea-facing site near Kinsale in County Cork. This 1-ha vineyard is planted entirely with Rondo, the original vines having been given to Walk by Professor Helmut Becker in 1982, and the wines can be very good. Many of the other Irish vineyards are also planted with Rondo as well as Solaris. Having visited a number of Irish vineyards in August 2024, I saw that these were the only two varieties showing what I would call reasonable yields. Quite a few of the vineyards are planted with a wide range of varieties – many the same as we are growing in Britain – but it is my view that only in extremely warm and dry years will these crop at a sustainable level, 2025 being such a year. However, as GB found in the 1950s and 1960s, you have to start somewhere and this is early days for Irish vineyards. Northern Ireland has no vineyards that I know of.

PIWIs in New Zealand

New Zealand vineyards, while dominated by Sauvignon blanc, Chardonnay and Pinot noir, also contain a few old hybrids and a few PWIs. None are of any real consequence today, although they were once. 4-ha of Chambourcin and 3-ha of Niagara, both old hybrids, seem to be the largest plantings today. However, the history of winegrowing in New Zealand could have been so different. I was present in 1982 at the 100th anniversary of the crossing by Dr Müller of the variety that came to be known as Müller-Thurgau. This was held at Geisenheim, where the crossing had been made, and where I had been a guest student in 1975-76. Professor Helmut Becker, then heading the research station, invited me to the event, a huge tasting and a multi-wine banquet, and I was GB's representative because MT accounted for about 50 per cent of all our vineyards.

In the afternoon of the day before the banquet, I happened to see what appeared to be a private party taking place on a lawn by the main administration block. There were probably 20-30 people there. The beers were flowing and a whole lamb (probably a New Zealand one) was being barbequed. Becker was in charge, and before long took control of the event. A man came forward, who I later learned was the Prime Minister of New Zealand, Sir Robert Muldoon, and after some short speeches, hands were shaken, and he was presented with a certificate. This, while not quite the Treaty of Waitangi, could have been a seminal document for the NZ wine industry. It turned out to be a gift, in perpetuity, of the plant breeder's rights to the variety Breidecker. This variety was a crossing between a Riesling x Sylvaner cross with Chancellor, the old Seibel cross. It had been produced by Becker's predecessor Professor Heinrich Birk in 1962, and named after a famous NZ winegrower, Heinrich Breidecker. Birk had also produced Reichensteiner and Schönburger among several other varieties.

Unknown to most of the rest of the world, NZ had no need of new grape varieties as growers in the South Island had started planting Sauvignon blanc in 1972 and by 1982 were just about ready to launch themselves onto the world market. The rest is viticultural history. Breidecker never got a look in and although some did get planted and there are still some vineyards in Canada planted with it, it remains an obscure and very minor variety. I have always wondered whether it might suit GB, but I do not believe anyone has ever tried it.

To summarise: Are the most widely planted PIWI varieties the best?

As has been shown in the first part of this chapter, there are many different PIWI and hybrid varieties that could be planted in European vineyards. By area planted, the most widely planted are the older ones, some of which are declining in popularity as newer, better varieties come along. As a guide for 'what to plant' the wisdom of crowds is often a good guide and with the exception of Regent and Roesler, I would happily recommend any of the top eleven varieties currently planted in Europe which account for 81 per cent of all plantings (See Table 22). There are also some up-and-coming varieties that I would also consider, especially the Calardis range.

Question 2: Do PIWIs produce as good a wine as *vinifera* varieties?

The simple answer is NO. The intrinsic problem with producing new varieties with disease resistance as their main priority is that in exchange for these undoubtedly valuable attributes, wine quality is sacrificed. In addition, the fact that all PIWI varieties are crossed and trialled in warmer climates than GB's means that one needs varieties that in their place of origin have lower acid levels and earlier ripening dates to counteract the tendency for grapes grown in marginal climates to suffer from high acid levels (especially high levels of malic acid), low natural alcohol levels and late ripening dates. This is why around 35 per cent of my list of PIWIs worth growing in GB have *Amurensis* parentage, plus there are six varieties with muscat parentage as muscat varieties tend to have lower acids – think Gewürztraminer, the Muscats, Torrontes, and Viognier.

PIWI International Wine Competition

There are several ways of gauging wine quality and they all have their flaws. It is actually quite difficult to get a representative sample of single varietal wines, especially from the cooler climates, and of course there are many different ways of making wine. How can you decide whether a wine from a certain variety, made conventionally and bottled without too much adornment would be better made as an orange wine or a vibrant pét-nat? Only time will tell. One way I decided might yield some interesting observations was looking at the results of the PIWI International Wine Competition results for the years 2019 and 2022, 2023 and 2024 and seeing which varieties had been most successful in winning high points and awards. There were entries from most of the major European wine producing countries, France, Germany, Italy, and Austria, and in addition there were entries from several smaller and less well-known ones: Belgium, Czech Republic, Great Britain, Netherlands, Poland, Slovakia, Sweden, and Switzerland. Over the four years of the competitions looked at, around 750 wines were judged covering 146 different PIWI varieties. Of these, 97 received no, or minimal points. The 49 that did score points are ranked by points awarded as shown in Table 31.

PIWI International Wine Competition results 2019 & 2022-24

Variety	Points	Date
Solaris	6,098	1975
Muscaris	5,884	1987
Souvignier Gris	5,601	1983
Johanniter	5,089	1968
Regent	4,300	1967
Cabernet Blanc	4,024	1991
Cabernet Cortis	1,779	1982
Hibernal	1,642	1944
Sauvignac	1,529	1991
Pinotin	1,237	1991

Note: Date refers to date of crossing

Table 31

Of course, this list has a historical bias as, by definition, the early PIWIs are the most widely planted with Solaris, the second oldest variety (after Hibernal) in pole position and no varieties created later than 1991, more than 35 years ago. Regent, one of the oldest varieties and with the largest planted area, comes a meagre fifth, and there are only two other reds on the list, Cabernet Cortis and Pinotin, seventh and tenth respectively. Not a great showing for the reds which is very telling. It is also interesting to note which varieties are NOT there: Cabaret noir and Divico for instance. It would appear that new red varieties either have lower quality, or that red wine drinkers are more selective about the wines they enjoy – or maybe a bit of both.

As far as WineGB competitions are concerned, one of the old hybrids – Seyval blanc – has performed consistently well in the national wine competition since the early years. The competition was started by the English Vineyards Association (as it was then) in 1974 and continues under WineGB to this day. Although the first winners were made from Müller-Thurgau (or Rivaner as some people preferred to call it), the first Seyval gold medal winner appeared in 1978 and the variety has continued to shine ever since, appearing in both still and sparkling wine categories, and still winning gold medals. The major PIWIs grown in GB, Solaris, Rondo, Regent, Phoenix and Orion have all won medals, mainly bronzes, a few silvers and even fewer golds. In recent years some of the newer PIWIs, Cabaret noir, Divico, and Muscaris have also won a few medals, though nothing outstanding. I guess it takes time for both growers and winemakers to get to grips with any new variety, when to pick it and how to treat it in the winery. The advent of wine styles such as natural wines, pét-nats, and col-fondos[52] have in some senses thrown a lifeline to new varietals with unfamiliar (and frankly often bizarre) names. Anyone contemplating buying a wine made in one of these newer wine styles is probably willing to try it whatever obscure grape variety it is made from!

To summarise: Do PIWIs produce as good a wine as *vinifera* varieties?

As has been explained above, while the quality of wine from PIWIs improves with each generation of new varieties and with gene editing and who knows what other techniques yet to be invented, there may come a day when the wine quality from these varieties will be the equivalent of that produced by the finest *viniferas*. Of course, the question should be: are yields high enough and is the wine from PIWI varieties good enough to sell in the right quantities and at the sort of prices needed to keep the vineyard viable? This is a question that each vineyard has to answer in the light of their own circumstances.

Question 3: Disease resistance: Can PIWIs be successfully grown without using any pesticides?[53]

You can of course grow any grape variety without resorting to pesticides, but the question is, can it be done successfully? Can one run a truly sustainable vineyard without pesticides, especially in a marginal climate? A vineyard that will pay the interest on the capital it has consumed plus the annual running costs? These are the real questions.

First some biology. Vineyard owners are not really in the business of grape production but rather of leaf production. If you remove all the grape flowers from your vine before flowering, you vine will grow perfectly well, perhaps even better as its been relieved of the burden of grapes to ripen into which it had expected to store some of the sugars and starches that its leaves, through the miracle of photosynthesis, produces. Try removing all the leaves however, and you will very quickly find that the vine struggles to live, let alone produce enough sugars to survive. If we neglect the production of leaves and allow them to

52 Pét-nats are sparkling wines produced via a single bottle fermentation and do not go through the riddling and disgorging process. Col-Fondo wines are in essence Champagne method wines that have also not gone through the riddling and disgorging process. Both wines are usually cloudy and sometimes still going through the secondary fermentation process.

53 The term 'pesticides' covers all plant protection materials applied to the vines, including those allowed under organic and biodynamic regulations, whether by spraying, dusting or soil applied. The term also included herbicides (weedkillers).

become diseased, covered in mildew and lacking enough chlorophyll in order to photosynthesise properly, then the vine cannot perform well. The vine needs to do two things: send sugars and starches to its crop, but also to its woody parts – the permanent wood, trunk and roots, as well as its canes, especially those destined to provide next year's fruiting wood whether it be canes or spurs. This is in order for it to survive the winter and emerge in the spring with enough stored energy to grow leaves so that they can get to the point where they are big enough to re-supply the mother-ship. Leaves then are what we are in the business of growing and while they are chemical marvels, they are tender and need looking after.

I quite often get a phone call or an email along the following lines. 'Hi there, I would like to plant a vineyard, but I do not want to have to spray it. What varieties do you recommend?' I then explain that even the hardiest, disease resistant varieties, need a little bit of TLC which includes a few sprays, if only with simple materials such as wettable sulphur and copper-containing Bordeaux mixture. I also then tell them that the hardiest varieties do not necessarily make the best wine, and the conversation then starts getting difficult. The truth is that in a climate such as GB's that can be wet during the growing season, even the best PIWIs benefit from some form of chemical assistance, if only 3-4 sprays of fairly benign materials. There have been several reports about some of the older PIWI varieties – Bianca, Johanniter, and Regent, all bred in the 1960s – that they have in some vineyards started to lose their resistance and, although still better than *viniferas,* are not as disease proof as they once were.

So, back to the question posed at the beginning of this section: Can PIWIs be grown without using pesticides?

I have discussed the question of spraying with several nursery owners and research staff at vine breeding stations and they agree that most of the PIWIs now planted need three to four sprays during the growing season, especially around flowering time and before bunch-close, in order to protect them and keep them healthy. Here is a selection of what they say.

Erhard Tutzer, Innovitis Vine Breeding Institute, Bolzano

The late Erhard Tutzer was involved with breeding PIWIs, not that he called them that in the 1960s when he started his nursery. He speaks with authority.

Why is it so difficult to breed good and resistant varieties?

Because plants adapt and can also lose their resistance mechanisms again. This is the reason why it is important to work with multiple resistances in natural crossbreeding. In the past, varieties were also authorised that had not been sufficiently tested. Their resistance proved to be insufficient after a short time. Many [growers] had to uproot them again and suffered financial losses as a result. These experiences have made growers and winegrowers much more cautious.

How can you prevent the vines from losing their resistance mechanisms?

You should never completely dispense with plant protection but also treat the PIWI varieties two to three times a year with sulphur and a very reduced dose of copper. The first time should be soon after bud break so that any mildew infestation from the previous year can be combated at the source. The second spraying should take place during the flowering phase, as the pinhead-sized berries are very sensitive and unprotected. In extremely wet years, a third treatment may also be necessary, but the PIWI varieties generally have sufficient defence mechanisms.

Sam Middleton from Vineworks in Vineyard Magazine July 2024

While the number of spray applications in the UK will be considerably fewer than for the commonly planted Champagne varieties, PIWIs will require the same level of attention in all other aspects -

pruning, canopy management, yield management and vineyard floor management. What they do offer are reduced spray costs, reduced risk of yield lost to disease, and in some cases, higher potential yields to begin with.

Ben Brown, agronomist with Agrii in Fruit & Vine Magazine September 2023

Although the number of sprays per year depends on the variety, Ben Brown, viticulture agronomist at Agrii who has designed a number of programmes for PIWIs, now recommends 4-6 sprays per season. Robb Merchant, of White Castle Vineyard in Abergavenny, said that this fits in with his experience as a PIWI grower. Challenges, on the other hand, include maintaining the fungal resistance. **Fungal resistance is not an absolute and it is not for ever.** Maintaining the disease resistance of PIWI varieties is a key issue as breeding a new grape variety is a 25–30-year process – compared to 7–8 years for a new wheat variety for example, Ben explained. Maintaining disease resistance is therefore something we need to consider when planning fungicide programmes.

A common misconception regarding PIWIs is their tolerance to all diseases. Depending on the variety, there may be loopholes for black rot and anthracnose, Ben added. While these are generally less of an issue now, especially with conventional spray programmes, they can cause a high level of crop and yield loss – so this is something to think about where resistance has not been bred into these hybrids. The main concern would be early season black rot up to *véraison*, Ben continued, recommending 3–4 sprays depending on rain pressure. Some PIWI varieties are bred to have loose clusters so they are more disease tolerant for *botrytis*, however any wounds can still be colonised by spores later, so a couple of late season *botrytis* sprays may also be needed.

Generally, PIWIs have moved from monogenic resistance – offering single gene resistance – to predominantly two genes that convey resistance to fungal diseases. With *botrytis*, rather than the genetic disease resistance seen with powdery mildew and downy mildew, it is more of a physiological or phenological resistance, Ben explained.

To summarise: Can PIWIs be grown successfully without using pesticides?

There are a multitude of factors to consider in controlling plant diseases and the variety of plant is but one of them. The climate, the site, the planting density, the trellising and the disease protection measures, including spraying pesticides if necessary, are all part of the armoury. Appropriate pruning and effective canopy management are probably the two easiest (and cheapest) disease management techniques available to all vine growers. But there comes a time when most vine growers, growing as they are a tender sugar-producing plant, need to give the plant a bit of help, and spraying with a suitable pesticide (typically a fungicide) will do that. The idea that a vine can 'learn' to resist pests and diseases by exposing it to them is, at least for the time being, a fantasy, despite what those at the far end of the biodynamic spectrum might preach. Resistance to diseases is something that humans can get by exposure to them – this of course is the basis of immunization – but plants cannot do this. Resistance has to be bred into plants through the introduction of resistant genes.

Growing disease resistant hybrids and PIWIs can make life easier, but as we have read, the received wisdom is that three to four sprays during the growing season will help the plant maintain that natural ability to fight diseases.

Genetically modified vines

The use of genetically modified (GM) varieties in some species is very widespread and it is estimated that there are over 200 mn-ha of species such as Soyabean, Maize, Cotton, Canola (Oilseed Rape), Sugar Beet, Rice and other less important crops. The USA, Brazil, Argentina, Canada and India are the largest growers of GM crops and they are sold worldwide and probably consumed or used in almost every country. For vines, there are currently no GM varieties allowed to be used for winemaking, despite the fact that research stations have been growing them for decades. I recall visiting Geilweilerhof in the 1980s and being shown a small fenced-off plot of vines that had been genetically modified by having some genes from barley inserted into the plants to see if that would control *botrytis* (which it did). They told me that they had even made some wine from the grapes, but under no circumstances was anyone allowed to drink it! They could analyse it, look at it, sniff it, but nothing else. However, this experimental vineyard, and several others around Europe with GM plots, have all been grubbed. The main reason that nobody was allowed to commercialise GM vines is that it is felt that it would damage the public perception of wine (even though of course wine contains a known killer – alcohol!).

The drive to reduce pesticide use is now a much more important topic than it was, and various research stations are now trialling vines that have been 'gene edited' to switch off the genes that cause the plant to be affected by Powdery Mildew. These appear to be successful, although at least one plot has been vandalised by those against progress of this sort. Personally, I think that this is the way forward and when vines with natural resistance to the major pests and diseases become a commercial reality, then they will be taken up with enthusiasm. Jamie Goode in his book *The New Viticulture* writes at length about the pros and cons of these vines and says 'the emergence of GM vines may just hasten the progress of soulless big brands [of wine], further severing the connection that wine has with the soil'. While this might be true on the one hand, maybe having vines that are genetically exactly the same would tend to highlight the differences between wines, thus showing the *terroir* effect even more? While we all like drinking high-end individual wines that show their place of origin and show the winegrowing and winemaking skills of their producers, most wine is pretty ordinary, much of it not very good and what has worked for tomato soup or baked beans or cola, might well work for wine? And if naturally disease resistant vine varieties can be grown with fewer pesticides, with less impact on the environment and at less cost, doesn't that make sense?

Chapter 9

PIWI and hybrid vine varieties for Great Britain

The decision to plant any grape variety, let alone a relatively new, possibly untried, variety, needs to be taken with great care as the penalties for 'getting it wrong' are huge, both financially and quite possibly reputationally. Who wants to be known for planting the latest 'super-vine' only to find that it is a beast to grow, ripens too late and – the last straw – the wine quality is decidedly lacking?

Warnings

None of the PIWIs currently on the market have been produced with reference to GB growing conditions and all of the research stations working on PIWIs are in warmer climates than GB's.

There have been no proper trials carried out with PIWIs in GB and therefore most of the data available on yields and performance (sugar and acid levels) must be used with care. Growers who have unsuccessful varieties tend not to report them.

There are 38 different PIWIs and hybrids being grown in GB (2024 WS data) covering 444-ha. See Table 32 for list of PIWIs and hybrids. Many of these are growing in the West, Wales and Midlands & North regions where growing conditions are more marginal than the warmer and drier east and south-east. This will have a negative effect upon yields, performance levels and quality.

Current situation of PIWIs and hybrids in GB

The current plantings of PIWIs and hybrids in GB is shown in Table 32 and consists of 38 different varieties, of which four are old hybrids (Seyval blanc, Marechal Foch, Cascade and Chambourcin) and thirty-four are modern PIWIs. The old hybrids, all developed in the first half of the 20th century, have fallen slightly in area planted over the last 35 years. In 1990 they were planted on 133-ha. In 2024 they are planted on 123-ha. During that same period, the total vineyard area has more than quadrupled. Seyval blanc, despite its being almost a century since it was crossed, is still the fifth most widely planted variety in GB and has proved itself both in this country and overseas as a valuable variety with a good resistance to both mildews. An additional 11-ha were planted in 2003-4. The other three have all fallen in area and will surely disappear over the next few years.

The thirty-four modern PIWIs, of which there are 320-ha, are headed by Solaris, crossed in 1975 and appearing on the GB variety lists for the first time in 2009 when 4.74-ha was recorded. This has expanded now to 117-ha, making it the sixth most widely planted variety in GB and my guess is that it will continue to expand and will soon overtake Seyval blanc. The next most widely planted PIWI is Rondo, introduced to GB in 1982 when its breeder, Professor Helmut Becker, gave 50 vines each to me, Karl-Heinz Johner at Lamberhurst and Ken Barlow at Adgestone. It was officially welcomed onto the GB list of 'Authorised' varieties, along with three relatively recent PIWIs, Orion, Phoenix and Regent, shortly afterwards. Plantings of Rondo, Phoenix and Orion first appeared on the GB variety lists in 1990, with Regent not making an appearance until 2006. By 2009, total plantings of these four varieties had grown to a total of 98-ha. The total area planted of these four varieties today is 257-ha. Most of the growth in these plantings has come from Solaris and Rondo, with the others expanding more slowly.

The rise in plantings of the other PIWIs started in 2012 with very small plantings of

PIWIs and Hybrids in Great Britain 2024 - Hectares							
Colour	**Type**	**Variety**	**2024**	**Colour**	**Type**	**Variety**	**2024**
white	Hybrid	Seyval Blanc	122.71	red	PIWI	Bolero	0.40
white	PIWI	Solaris	117.07	white	PIWI	Zalagyongye	0.40
red	PIWI	Rondo (GM 6494/5)	64.23	red	Hybrid	Marechal Foch	0.35
red	PIWI	Regent	32.42	red	PIWI	Artaban	0.30
white	PIWI	Phoenix	28.95	white	PIWI	Fleurtai	0.30
red	PIWI	Caberet Noir	14.75	white	PIWI	Soreli	0.30
red	PIWI	Divico	14.03	white	PIWI	Kersus	0.22
white	PIWI	Orion	13.96	white	PIWI	Calardis Musque	0.20
red	PIWI	Cabernet Cortis	6.62	red	PIWI	Cabernet Jura	0.16
white	PIWI	Sauvignac	6.04	white	PIWI	Hibernal	0.16
white	PIWI	Muscaris	5.67	white	PIWI	Floréal	0.15
white	PIWI	Souvignier Gris	4.61	red	PIWI	Pinot Kors	0.14
red	PIWI	Pinotin	3.58	red	Hybrid	Cascade	0.13
white	PIWI	Johanniter	1.48	red	PIWI	Cabernet Cantor	0.12
white	PIWI	Cabernet Blanc	1.10	white	PIWI	Saphira	0.07
white	PIWI	Voltis	1.02	white	PIWI	Ravel blanc (VB-37)	0.05
white	PIWI	Villaris	0.75	white	PIWI	Pölöskei Muskotály	0.01
white	PIWI	Calardis Blanc	0.67	red	PIWI	Muscat bleu	0.01
red	PIWI	Bronner	0.55	red	Hybrid	Chambourcin	0.01
Total hectares							**443.69**
Source: Wine Team at the Food Standards Agency 2025							

Table 32

PIWIS & Hybrids in GB vineyards - 2024			
Variety	**Number of vineyard parcels**	**Variety**	**Number of vineyard parcels**
Seyval Blanc	287	Divico	34
Rondo	246	Muscaris	34
Solaris	220	Sauvignac (VB Cal 6-04)	14
Regent	148	Souvignier Gris	13
Phoenix	131	Pinotin	8
Orion	59	Johanniter	8
Cabaret Noir	37	Cabernet blanc	7
Cabernet Cortis	36	Voltis	4
Source: Wine Team at the Food Standards Agency			

Table 33

Bronner and Bolero, expanding in 2015 with some Cabernet Cortis. The appearance of Divico on the lists in 2022 was almost entirely due to some publicity given to it by the person then running the experimental vineyard at East Malling Research Station (more properly called the National Institute of Agricultural Botany or NIAB-EMR) who sang the praises of the variety having planted a block of it in 2016. It was said in a press release that 'Divico is likely to supersede

Pinot noir as the industry standard red variety'. Most of the Divico at East Malling has now been grubbed out as it was found to be difficult to ripen. Meanwhile, quite a few growers took heart at what they heard from East Malling and the area of Divico rose from 4.50-ha in 2022 to 14.03-ha in 2024. Whether or not Divico will continue this upward trajectory remains to be seen. The wines have not yet excelled (although the very warm growing conditions in 2025 might change that). Growers have continued to plant PIWIs, with the area rising by 40-ha (9 per cent) in 2024. I can foresee this rising still further, although to put it in context, the area of *viniferas* rose by 11 per cent over the same period. The number of vineyard 'parcels' (separate plots of land planted with vines) is shown in Table 33 and gives a fair idea of the popularity of different PIWI and hybrid varieties.

Increase in plantings of PIWIs in cooler regions of GB

The increase of plantings of PIWIs has been strongest in the cooler and more marginal regions of GB. As can be seen from Table 34, plantings of PIWIs and hybrids in Wales, the Midlands & North, and the West regions are all in double digits in percentage terms, with Wales and the Midlands & North both being over 50 per cent of total vineyard area. My worry is that growers in these regions are planting them in the belief that they can be grown without, or at least with fewer, pesticide applications, that they are naturally high-yielding and that they are more suitable for a cooler and wetter climate. It is my opinion that none of these three factors is necessarily true. As I said at the beginning of this chapter, none of the PIWIs currently available have been produced with reference to GB growing conditions and all of the research stations working on PIWIs are in warmer climates than GB's.

Recommended PIWI and hybrid vine varieties for vineyards in Great Britain

The PIWIs and hybrid varieties shown in Table 35 are those that, depending on site quality, I consider suitable for planting in GB. However, they will not suit all parts of the country. Those growers with exposed sites, sites at altitudes of 120 m above sea level or higher, in parts of the country that have low average GDDs (800 or below) and a low average number of days of 30°C or more, should select low-acid, early ripening varieties.[54] The production of dark, robust red wines always requires good heat levels and on marginal sites will prove difficult. Varieties for

54 It is important to note the word 'average' in these requirements. As we have seen, in our maritime climate there can be good years (2018 and 2025) and bad years (2012 and 2024) and varieties have to be selected with these fluctuations in mind. Selecting a variety that only does well in good years will not pay the bills.

Distribution of PIWIs & hybrids in GB regions

Region	Hectares planted	Viniferas	PIWIs & Hybrids	% PIWIs & hybrids
Wales	91.35	41.48	49.87	54.6%
Midlands & North	195.82	95.89	99.93	51.0%
West	377.90	275.42	102.48	27.1%
T&CVA	138.20	127.96	10.24	7.4%
East Anglia	673.01	631.25	41.76	6.2%
Wessex	594.21	566.07	28.14	4.7%
South East	2,510.64	2,405.52	105.12	4.2%
Total	**4,581.13**	**4,143.59**	**437.54**	**9.6%**

Source: Wine Team at the Food Standards Agency

Table 34

PIWIs and hybrids for Great Britain					
Variety Name	**Ha - 2024**	**Date of crossing**	**Variety Name**	**Ha - 2024**	**Date of crossing**
White varieties			**White varieties - continued**		
Blütenmuskateller	0.00	1947	Sauvignon Rytos	0.00	2002
Cabernet Blanc	1.10	1991	Seyval blanc	122.71	1930s
Calardis Blanc	0.67	1993	Solaris	117.07	1975
Calardis Musqué	0.20	1964	Soreli	0.30	2002
Calardis Soleil	0.00	2010	Souvignier Gris	4.61	1983
Donauriesling	0.00	1978	Voltis	1.02	2002
Fleurtai	0.30	2002	**Red Varieties**		
Johanniter	1.48	1968	Cabaret noir	14.75	1991
Muscaris	5.67	1964	Cabernet Cortis	6.57	1982
Orion	13.96	1987	Cabertin	0.00	1991
Phoenix	28.95	1964	Cal 1-28	0.00	1991?
Ravel Blanc (Sauvignon Soyhières)	0.05	1998	Divico	14.03	1997
Réselle (Recelle)	0.00	1981?	Pinotin	3.58	1991
Riesel	0.00	1989	Prior	0.00	1987
Rinót	0.00	Not known	Regent	32.42	1967
Sauvignac	6.04	1991	Rondo	64.23	1964

Source: Wine Team at the Food Standards Agency

Table 35

making red wines must therefore be chosen with extra care.

The thirty-one varieties listed above include some which are not (yet) planted in GB. Those that are currently planted occupy 439.71-ha which equates to 9.45 per cent of the total GB vineyard area of 4,650.90-ha.

Great planting weather

Valentin Blattner

Valentin Blattner lives in Switzerland in the village of Soyhières near Basle where he owns a vine nursery and vineyards. He is the PIWI producer extraordinaire and ten of his varieties are among the top forty – 23 percent – grown in Switzerland, the country with one of the highest penetrations of non-viniferas. Ten of the thirty-one varieties I list below as being suitable for GB are Blattner crossings, an even higher percentage.

Many of Blattner's varieties stem from crossings he produced in 1991 using Cabernet Sauvignon as one parent and various PIWIs and hybrids as the other. He works closely with Volker Freytag from Rebschule Freytag near Neustadt in Germany who has taken over the commercialisation of his varieties. Six of the seven Freytag 'Pioneer Wines' are Blattner crossings.

www.domaineblattner.ch

The varieties are listed in alphabetical order.

Blütenmuskateller

Type: PIWI
Colour: White
Origin: Potapenko All-Russia Research Institute, Rostov-on-Don, Russia
Crossing: Severnyi x (Mixed muscat pollen)
Date of crossing: 1947
Year of registration: Not known
Planted area in Austria: 124-ha
Planted area in GB (2024): None

Blütenmuskateller is a crossing made at the Potapenko All-Russia Research Institute near Rostov-on-Don in Russia, the same institute that produced some of the *Amurensis* crossings grown in GB such as Rondo. Despite being a relatively old crossing (made in 1947), it has good natural disease resistance against Downy Mildew but requires some sulphur sprays to keep Powdery at bay. Good leaf removal is recommended. With 124-ha, it is the second most widely grown PIWI in Austria, although it is not grown in Germany at all. The variety ripens after Pinot noir but before Chardonnay and wine quality is good with a classic muscat tone and balanced acidity. Natural sugars can be towards the lower end of the spectrum. It is ideal for blending with more neutral varieties.

Cabaret noir[55]

Type: PIWI
Colour: Red
Origin: Blattner, Valentin
Crossing: Cabernet Sauvignon x Unknown resistance partners
Date of crossing: 1991
Year of registration: Not known
Planted area in Belgium (2023): 8.16-ha
Planted area in GB (2024): 14.75-ha

Information on this variety is scarce but Volker Freytag who has 'taken over the protection and propagation' of the Blattner varieties states:

"The cluster is short, discreetly loosened, with small shoulders. The berry is round, small and covered with an even layer of wax. The thick berry skin reduces the risk of *botrytis*, which has a positive effect on the harvest period. The ripening time is early, the beginning of September (in Germany) which makes the variety suitable for late or high-altitude sites or more northerly (European) wine regions The growth is medium strong and upright. Due to the short internodes, the grapes are all in an even horizon, which makes the work of defoliation much easier when necessary. It is prone to damage from SWD.[56]

The wine shows a subtle Cabernet play in the aromas of dark cherry, juniper berry and violet. On the palate, it is reminiscent of the spice of cloves and pepper. The ripe tannins round off the pleasantly soft flavour profile of this variety."

This variety has not been planted very much in other European countries (and is not planted at all in Germany). It would appear that GB has the largest area under cultivation in the world with 37 separate vineyard parcels planted. Essex has the largest planted area with 2.68-ha. Artelium Vineyard in East Sussex and The Wharie Experience in Hampshire both have wines made from it which are selling at £30 to £35 a bottle. This might be an interesting variety, but I would plant it with caution and, as with all varieties for red wine, only on a good site. This variety is one of Volker Freytag's seven 'Pioneer Wines' (his name for PIWIs).

55 The name Cabernet noir was originally chosen for this variety when it was released in Switzerland (where it was developed), but the EU objected so Cabaret noir was chosen instead.

56 SWD is Spotted Wing Drosophila or Japanese Fruit Fly

Cabernet blanc

Type: PIWI
Colour: White
Origin: Blattner, Valentin
Crossing: Cabernet Sauvignon x Regent
Date of crossing: 1991
Year of registration: 2008
Planted area in Germany (2024): 322-ha
Planted area in GB (2024): 1.10-ha

Cabernet blanc is Germany's third most widely grown PIWI, and there are plantings in Austria and Belgium (and one whole hectare in GB). It is said to be hardy in winter frosts, with an upright growth habit and relatively few leaves in the grape zone which ensures good ventilation. It has some natural resistance against Powdery Mildew, less good against Downy, and good *botrytis* resistance. It will benefit from a basic spray programme.

The one major issue that everyone writes about is that in some situations it can suffer badly from *coulure* or *millerandage*, thus reducing the crop and making crops uneven from year to year. This is similar to Huxelrebe which also suffers from *millerandage* although the unfertilised grapes will become very sweet. Wasps can be a problem. It ripens quite late, with or just before Chardonnay. However, it is on my recommended list because of its excellent wine quality in a Sauvignon style. In the PIWI International Wine Competitions it received very high scores thus indicating the excellent quality of the variety. For GB growers I recommend it for good to very good sites, ones where Chardonnay will ripen successfully. Do not expect huge yields, but the wine should be good.

Cabernet Cortis

Type: PIWI
Colour: Red
Origin: Freiburg
Crossing: Cabernet Sauvignon x Solaris
Date of crossing: 1982
Year of registration: 2004
Planted area in Germany (2024): 68-ha
Planted area in GB (2024): 6.57-ha

Cabernet Cortis is one of the considerable number of crosses produced by Dr Norbert Becker during his time as the head of the Freiburg vine breeding station. It has achieved modest coverage in Germany, is the second most widely planted red PIWI in Belgium, and is to be found in quite a number of other countries including even France (50-ha). It is said to grow upright with few side shoots, and from its *Amurensis* parentage, it gets its early ripening. It is said to ripen 7 days before Pinot noir. It has good to medium-good resistance to both mildews and *botrytis* but will benefit from a basic spray programme. It is planted in 37 vineyard parcels in GB, with very few in the south-east. Like most red varieties, especially in early vintages, it is prone to damage from SWD.

Cabernet Cortis produces wines with pronounced Cabernet Sauvignon characteristics, and its wines were the second highest scoring red varietal in the PIWI International Wine Competitions. It is on the basis of its wine quality that it is recommended. It needs a good site in GB and is not a variety for cooler regions. It needs to be fully ripe in order to avoid greenness in the wine.

Cabertin

Type: PIWI
Colour: Red
Origin: Blattner, Valentin
Crossing: Cabernet Sauvignon x Regent
Date of crossing: 1991
Year of registration: 2008
Planted area in Germany (2023): 24-ha
Planted area in GB (2024): None

Cabertin is a Blattner cross and one which does not use his traditional 'resistance partners' but instead uses Regent as father which has both Villard and Seibel varieties in it. It ripens about 10-14 days earlier than Cabernet Sauvignon which means quite late in GB terms. Of all the Cabernet-based crosses, this one is the best in terms of wine quality with pronounced Cabernet Sauvignon characteristics. Its growth habit is upright, with an open canopy, and its resistance against Powdery Mildew and *botrytis* is very good, and good against Downy Mildew. It has a thick skin which is both good for *botrytis* control and wine colour. It will need a basic spray programme. It must be grown on a very good site and is NOT for cooler parts of the country.

Cal 1-28

Type: PIWI
Colour: Red
Origin: Blattner, Valentin
Crossing: Said to be a Muscat crossing
Date of crossing: 1991
Year of registration: Not known
Planted area in Germany (2023): None
Planted area in GB (2024): None
Planted area in Switzerland (2023): 10.66-ha

Cal 1-28 (sometimes called VB Cal 1-28) is another Blattner crossing that is finding favour in some areas. Switzerland seems to have taken to it, and there are several producers who are making wines and achieving success in competitions and on the market. Most of these wines are heavy red ones, and all of them have been aged for 12-24 months in *barriques*. This suggests a variety that has a lot of tannin and therefore needs a very good site to ripen it. Not for cool regions. It is said to have 'good to very good resistance to mildews and *botrytis* with small to medium grapes and a loose bunch.' This is a variety to plant with caution, very much as an experiment. Someone needs to come up with a good name for it. At the moment it sounds like a train time.

Calardis blanc

Type: PIWI
Colour: White
Origin: Geilweilerhof
Crossing: Calardis Musqué x Seyve-Villard 39-639
Date of crossing:1993
Year of registration: 2018
Planted area in Germany (2023): 90-ha (Est)
Planted area in GB (2024): 0.67-ha

Calardis blanc is a crossing of Calardis Musqué (which is Bacchus x Seyval blanc) with another Seyve-Villard crossing in order to improve its resistance against Downy Mildew. This appears to have worked because it has one more resistance gene. It also has two resistant genes against Powdery and one against Black Rot, and its disease resistance is rated as 'middling good.' It also has a loose bunch structure which helps against *botrytis*. It will benefit from a basic spray programme.

It is said to have an upright growth habit with a loose leaf-wall and yields are said to be good. Wine quality is also good with high sugars and low acids (in Germany). In GB conditions it should ripen just after Bacchus and before Pinot noir and Chardonnay. Acid levels are moderate, and the wine is good for both still and sparkling. It isa said to have Riesling characteristics. Calardis blanc was only released in 2020, so in viticultural terms, yesterday. Expect to see more of it.

Calardis Musqué

Type: PIWI
Colour: White
Origin: Geilweilerhof
Crossing: Bacchus x Seyval blanc
Date of crossing: 1964
Year of registration: 2023
Planted area in Germany (2023): None
Planted area in GB (2024): 0.20-ha

If there was ever a variety that could be said to be suitable for GB, one would have to say that a variety with Bacchus as one parent, and Seyval blanc as the other would be it. However, we all know that children never quite have all the attributes of both parents, and so it is with vines. It has a 'medium' resistance to both mildews and its loose bunch structure helps against *botrytis*. Yields are said to be low (by German standards) so probably not that low.

Calardis Musqué is loose-bunched and early ripening, both of which are welcome attributes in GB. Sugars are said to be higher than Müller-Thurgau and acids about the same. As with Bacchus, the Germans say this is suitable for making sweet wines, but in GB, I think it would be an ideal for dry to off-dry wines and as a blending partner with other less-flavourful varieties. If you wanted to grow Phoenix (another Bacchus x Seyve-Villard cross), then it would be an ideal partner. I would not bet the farm on it, but as an interesting variety to add to the overall *encépagement* this one could be useful.

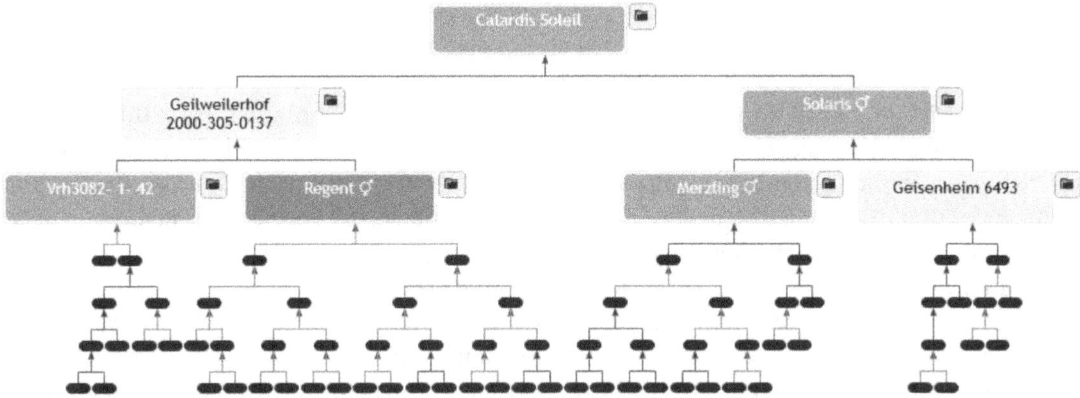

Calardis Soleil Family Tree

Calardis Soleil

Type: PIWI
Colour: White
Origin: Geilweilerhof
Crossing: Gf 2000-305-0137 x Solaris
Date of crossing: 2010
Year of registration: 2024
Planted area in Germany (2023): None
Planted area in GB (2024): None

Calardis Soleil is the latest PIWI to be released from Geilweilerhof, and benefits from the most recent developments in vine breeding. It has both Regent and Solaris in its parentage which gives it *Amurensis* and muscat in its breeding line, plus also Seyval blanc. With 5 genes against powdery and downy mildews and black rot, it has amazingly good disease protection and is undoubtedly the most disease resistant of all the modern PIWIs. The literature says that it does not suffer from sunburn, maybe an advantage if our summers get much warmer.

Bunches are loose and grapes mainly small. Yields should be good to very good. It is an early variety and in GB should ripen just after Solaris i.e. with or slightly before Bacchus. Its wine quality is excellent, and it should ripen easily to give full-bodied, lightly fruity wines, with ripe acidity. Geilweilerhof say that there should be an '80 per cent reduction in pesticides'. Definitely one to watch out for.

Divico

Type: PIWI
Colour: Red
Origin: Agroscope Research Centre, Pully, Switzerland
Crossing: Gamaret x Bronner
Date of crossing: 1991
Year of registration: 2008
Planted area in Switzerland (2023): 94.93-ha
Planted area in GB (2024): 14.03-ha

Divico comes from the Agroscope Research Centre in Pully, Switzerland (the same place that sent Müller-Thurgau to Ray Brock in 1947) and is Gamaret x Bronner. Gamaret is Gamay x Reichensteiner and Bronner is Merzling x Gm 6494. The Gm vine is a *V. Amurensis* cross from the breeding programme that produced Rondo. Divico has been taken up in quite a big way by the Swiss who now grow 95-ha of it, making it their largest PIWI variety by far.

The appearance of Divico on the lists of varieties being grown in GB in 2022 was almost entirely down to some publicity given to it by the person then running the experimental vineyard at East Malling Research Station (now more properly called the National Institute of Agricultural Botany or NIAB-EMR) who sang the praises of the variety having planted a block of it in 2016. He said in a press release that 'the Divico grape could be the much sought after game-changer' for red wine production in Great Britain as it produces 'quality red wine and thrives despite the challenges of the British climate.' He continued: 'Divico is likely to supersede Pinot noir as the industry standard red variety.' Time will tell whether this is right, not quite right, or not right at all. However, the new person in charge of the vineyard at East Malling has now grubbed up most of the Divico as it was found to be too difficult to ripen, so not a great vote of confidence.

Meanwhile, quite a few growers took heart at what they had heard from East Malling and the area of Divico rose from 4.50-ha in 2022 to 14.03-ha in 2024. It is planted on 34 vineyard parcels in GB, spread fairly evenly around the country. Whether Divico will continue this upward trajectory remains to be seen. The wines have not (yet) excelled (although maybe the über-warm conditions in 2025 will help it prove otherwise). The variety is late ripening and the quality very vintage dependent. Plant with caution is my best advice and only on your best sites.

Donauriesling

Type: PIWI
Colour: White
Origin: Klosterneuburg, Austria
Crossing: Riesling x Freiburg 589-54
Date of crossing: 1978
Year of registration: Not known
Planted area in Austria (2023): 81-ha
Planted area in GB (2024): None

Donauriesling is a complex hybrid from the Austrian research station in Klosterneuburg. It is Riesling crossed with a variety called Freiburg 589-54 which is Seyve-Villard 12-481 × Freiburg 153- 39 (the last of which is Pinot Gris × White Gutedel). It is the fourth most widely planted PIWI in Austria. The variety has a semi-erect to erect growth habit, with small berries and a loose bunch. Disease resistance is moderate, and it will need a basic spray programme. It is a late ripening variety (with Chardonnay) and requires a good site as acids can be high. However, the wine quality is excellent, probably the closest of all PIWIs to Riesling, and it is on this basis that it is recommended. Probably better for sparkling wines than still, although with a good site and in a ripe vintage, still wines can be made.

Fleurtai

Type: PIWI
Colour: White
Origin: University of Udine and Institute of Applied Genetics (IGA)
Crossing: Tocai Friulano x Kozma 20-3
Date of crossing: 2002
Year of registration: Not known
Planted area in Germany (2023): None
Planted area in GB (2024): 0.30-ha

Fleurtai is a relatively new variety which Vivai Cooperativi Rauscedo (VCR) have in their stable of PIWIs. Kozma 20-3 is a Hungarian crossing of Bianca (a complex Muscat crossing) with another Hungarian cross with both muscat and *Amurensis* parentage. It has 2 genes for resistance to both Powdery Mildew and one gene against Downy and overall its disease resistance is said to be 'excellent'. It also has good winter hardiness. The *Amurensis* parentage accounts for its early ripening and low acidity, both good for GB conditions. Its growth habit is said to be 'semi-upright or drooping' so will need good canopy management and timely leaf removal. The skin is said to be thin so be aware of *botrytis* issues. Its inclusion in my recommended list is based on its very good wine quality. It is a sibling to the variety called Soreli.

Johanniter

Type: PIWI
Colour: White
Origin: Freiburg
Crossing: Riesling x (Seyve-Villard 12-481 x (Ruländer x Gutedel))
Date of crossing: 1968
Year of registration: 1997

Planted area in Germany (2024): 141-ha
Planted area in GB (2024): 1.48-ha

Johanniter is one of the older PIWIs that is still being planted and one that is found in all of the northern European countries. It is Belgium's most widely planted PIWI although in GB it is only planted in 8 vineyard parcels. It has Riesling as one parent, with a Seyve-Villard cross, and Pinot gris (Ruländer) and Gutedel (Chasselas) in the other half.

As an older crossing, it is not as disease resistant as of some of the more modern crosses, although it has a two gene defence against Powdery and one against Downy Mildew. A minimum spray programme is advised. It has an early budburst, so avoid frost-prone sites and clusters are said to be compact. It is also quite a late ripening variety, with Chardonnay, and acids can be high. It needs to be planted on high quality sites in a warmer part of GB, not on marginal sites. Its inclusion in my list of 'recommended' varieties is only down to its evident wine quality, shown by the high number of points awarded in the PIWI International Wine Competitions surveyed. Growers interested in its wine quality should be able to find wines from both GB and other cool regions so that they can make a personal assessment of its style and quality.

Muscaris

Type: PIWI
Colour: White
Origin: Freiburg
Crossing: Solaris x Gelbe Muskateller (Yellow Muscat)
Date of crossing: 1987
Year of registration: 2012
Planted area in Germany (2023): 166-ha
Planted area in GB (2024): 5.67-ha
Planted area in Austria (2023-24): 97-ha
Planted area in Belgium (2023): 23-ha

Muscaris is a complex crossing produced at Freiburg by Dr Norbert Becker. It has both muscat and *Amurensis* parentage, which bring earliness and low acidity to the variety, plus Seyval blanc for natural disease resistance, all welcome in marginal climates. Resistance to both mildews is good, but not as good as more modern PIWIs and it will need a basic spray programme. It can get short of Magnesium, so foliar feeds may be required. Some reports say that it grows 'like a bush' so shoot selection pre-flowering, and good canopy management is advised. It is said to have an early budburst, so avoid spring frost-prone sites. It ripens relatively early, around the same time as Müller-Thurgau and can have high natural sugars. The bunch is said to be 'loose to very loose'. With the second highest points scored in the PIWI International Wine Competitions, Muscaris is a must for anyone wanting great wine quality. In some years, acids may be high, so good for sparkling wine.

It is to be found in 34 vineyard parcels in GB, mainly away from the south-east. There is none for instance planted in Essex, Kent, or the Sussexes. Growers interested in its wine quality should be able to find wines from both GB and other cool regions so that they can make a personal assessment of its style and quality.

Orion

Type: PIWI
Colour: White
Origin: Geilweilerhof
Crossing: Optima x Seyve-Villard 12-375 (Villard blanc)
Date of crossing: 1964
Year of registration: 1994
Planted area in Germany (2023): None
Planted area in GB (2024): 13.96-ha

Orion is one of the many crossings made by Professor Dr Alleweldt at Geilweilerhof. It dates from 1964 and was first registered in 1994. It is a crossing of another Geilweilerhof crossing,

Optima, which is ((Silvaner x Riesling) x MT), and Seyve-Villard 12-375 which is known as

Villard blanc. It is not as resistant to fungal

attack as some of the older hybrids (Seyval blanc for instance) and usually requires spraying against Powdery Mildew but is good against *botrytis* and ripens its wood well. It also shows better resistance to winter frost damage than MT (not often a problem in Britain) and its basal buds are said to be very fruitful.

Orion was introduced to Britain in the late 1990s and expanded very slowly and got to 10.8-ha in 2009. After that it went up and down a bit until it expanded in 2024 to almost 14-ha and planted in 59 vineyard parcels. Wines made from 100 per cent Orion seldom surface, so it is difficult to judge its true worth, but the wine is said to be fruity and aromatic. I do not see the area planted with it expanding much more, although some of the more northerly vineyards seem to like it.

Phoenix

Type: PIWI
Colour: White
Origin: Geilweilerhof
Crossing: Bacchus x Villard blanc (Seyve-Villard 12-375)
Date of crossing: 1964
Year of registration: 1992
Planted area in Germany (2024): 39-ha
Planted area in GB (2024): 28.95-ha

Phoenix is another of the many crossings made by Professor Dr Alleweldt at Geilweilerhof and is a Bacchus x Seyve-Villard 12-375 (Villard blanc) cross. The varieties Sirius and Staufer are siblings. Although not as resistant to fungal attack as the older hybrids, it has some natural defence, and it ripens its wood well.

It first appeared in Britain in the late 1980s and has risen very slowly from 20-ha in 2009 to 29-ha in 2024. It is planted on over 130 vineyard parcels, almost all in the west, or midlands and north regions, mainly because Three Choirs, the biggest winery in the west of the country, planted it in around 2002 and have been successful with it over the years. It achieves higher sugars and lower acids than Müller-Thurgau and on the palate is Bacchus-like, although not as powerful. If you wanted to plant a PIWI to give you wines in the style of Bacchus, then Phoenix is the one. To get the best wines, it needs a good site, and it will need a minimal spray programme to keep yields up. Growers interested in its wine quality will be able to find wines from both GB and other cool regions so that they can make a personal assessment of its style and quality.

Pinotin

Type: PIWI
Colour: Red
Origin: Blattner, Valentin
Crossing: Cabernet Sauvignon x Regent
Date of crossing: 1991
Year of registration: 2007
Planted area in Germany (2024): 48-ha
Planted area in GB (2024): 3.58-ha
Planted area in Belgium (2023): 7.23-ha

Initially thought to be a Pinot noir crossing (hence the name), DNA analysis has shown that Pinotin is in fact a crossing of Cabernet Sauvignon, from which it gets its wine character and quality and Regent from which it gets its Seibel and Seyve parentage which confers the natural resistance to diseases. It has very good protection against both mildews and *botrytis* and it has moderate growth which tends towards the upright, both aiding crop health. It is said it has a late budburst, so good for frost prone blocks. It ripens early (for a Cabernet Sauvignon crossing) and is said to ripen easily. Acids are said to be 'low' which probably means medium in GB terms. This variety is one of Volker Freytag's seven 'Pioneer Wines'.

Pinotin has achieved a moderate distribution around Europe and is becoming better known on account of the quality of its wine. While it only came 10th in the list of wines that were awarded the most points in the PIWI International Wine Competitions surveyed for this publication, given that there were 146 different varieties in the competition, that's not a bad achievement. It has something of a following in GB, being planted on eight sites, including some in Hampshire and

Essex. Like all robust red varieties, it needs a really good site in GB and will not produce great red wine every vintage.

Prior

Type: PIWI
Colour: Red
Origin: Freiburg
Crossing: Freiburg 4-61 x Freiburg 236-75 R
Date of crossing: 1987
Year of registration: 2004
Planted area in Germany (2024): 25-ha
Planted area in GB (2024): None

A complex Freiburg cross with Pinot noir on one side and a Geisenheim *Amurensis* cross (a Rondo sibling) on the other.[57] This combination gives the variety wine quality, plus very good mildew and *botrytis* defences, attributes that make it ideal for GB vineyards. It has strong, upright growth which keeps the canopy open, thus aiding disease control. It is said that the robust skin tends to 'disintegrate relatively quickly' once it approaches ripeness, so 'keep an eye on the harvest date'. Like most red varieties, especially in early vintages, it could be prone to damage from SWD. It ripens late, after Pinot noir, so needs a very good site to make the best wines. Definitely worth trying.

Ravel blanc

Synonym: Sauvignon Soyhières, VB-32-07 (or VB-32-7)
Type: PIWI
Colour: White
Origin: Blattner, Valentin
Crossing: Cabernet Sauvignon x *Vitis Amurensis Ruprecht*
Date of crossing: 1990
Year of registration: Not known
Planted area in Germany (2024): None
Planted area in GB (2024): 0.05-ha

Ravel blanc, the name the breeder prefers, is also confusingly known as Sauvignon Soyhières (or sometimes even Cabernet Soyhières), is a white, fruity variety, crossed by Blattner in 1990. It has a sibling called 'Petit Milo' which is grown in the cooler regions of Canada. Ravel blanc has good *Amurensis* genes which bring earliness and a high ripening potential and also good resistance against mildews and *botrytis*. The grapes are small, the bunch loose and the skins are thick. The grapes turn pink to light brown towards harvest and can darken considerably if left to hang.

As the name suggests, the wine has a light Sauvignon character which strengthens as the natural sugar level rises. Volker Freytag, the nurseryman that grafts more PIWIs than any other German nursery and has worked closely with Blattner over thirty years, rates it highly, and under the name 'VB32-7' it was chosen as one of seven of his 'Pioneer Wines'. I have tasted the wine while staying with the Blattners and can 100 per cent recommend it. I believe it to be one of the best PIWIs for GB and better than Sauvignon blanc if you want that style of wine.

Regent

Type: PIWI
Colour: Red
Origin: Geilweilerhof
Crossing: Diana x Chambourcin
Date of crossing: 1967
Year of registration: 1994
Planted area in Germany (2024): 1,525-ha
Planted area in GB (2024): 32.42-ha

Regent (pronounced in German *Ray-ghent* with a hard 'g' as in Ghent, not as in Regent Street) is another of Professor Dr Alleweldt's Geilweilerhof crossings (see Orion and Phoenix). Its parents are Diana, a Silvaner x Müller-Thurgau crossing (and not the native Catawba variety grown in the USA of the same name) and Chambourcin, a Joannès Seyve crossing of Seyve-Villard 12-417 and Plantet, an old Seibel crossing. According to *Wine Grapes* it is named after Le Régent, the 410-carat diamond found by a slave in the Kollur mine at

57 Freiburg 4-61 is Joannes Seyve 234-16 x Pinot noir and Freiburg 236-75 R is Merzling x (Zarya Severa x St Laurent)

GB - Yields of major red vine varieties 2016-2024 - Tonnes per hectare

Tonnes per active Ha	2016	2017	2018	2019	2020	2021	2022	2023	2024	Average t-ha	Average t-acre
Früburgunder, Blauer	1.81	3.50	6.62	4.32	3.55	5.34	5.59	6.97	1.80	4.39	1.78
Meunier	5.88	5.24	9.53	6.23	5.36	4.66	6.56	10.49	3.50	6.38	2.58
Pinot noir	3.00	4.31	7.93	6.18	5.25	5.70	6.01	9.20	3.90	5.72	2.31
Regent	No data	5.70	11.09	4.12	3.10	6.25	8.09	9.94	2.50	6.35	2.57
Rondo	No data	9.24	7.30	2.66	5.17	6.09	7.01	9.18	2.60	6.16	2.49
Averages	**3.56**	**5.60**	**8.49**	**4.70**	**4.49**	**5.61**	**6.65**	**9.16**	**2.86**	**5.80**	**2.35**

Source: Data collected by S. P. Skelton from growers (2016-17) and data collected for the WineGB annual yield report (2018-24)

Table 36

Golkonda in India and eventually placed in the crown of the French King Louis XV when he was crowned in 1722.[58]

The Regent crossing was made in 1967 and in Germany it was released in 1994. It was the first of the new generation of PIWIs and it has become one of Germany's leading varieties. It is also a variety that is grown in almost all other cool regions and for those that want more disease resistant varieties. However, being first has its disadvantages and others have you in their sights. Today (2025) other red varieties are starting to steal some of Regent's thunder, and as vineyards planted with this variety in the late '90s and early '00s start to age, they are being replaced by other, even better, red varieties. In 2008 Germany had 2,161-ha of Regent, the highest area it ever achieved, but today it stands at 1,565-ha and falling, although it is still their seventh most widely grown red vine variety. In 1996, despite not being 100 per cent *vinifera,* it was authorised for the production of Quality Wines in Germany, the first modern interspecific cross to be allowed to do so.

In Britain, Regent has been grown since the early 1990s and by 2006 there were 12.6-ha. It has expanded slowly to 32.42-ha in 2024 and is the third most widely planted purely red wine variety (i.e. ignoring Pinot noir and Meunier). It requires a good site, good management and deleafing at the right times, but no more so than other varieties for red wine production. It is certainly less disease-prone than Dornfelder and Rondo, so on that score alone it could be a better bet for a deep red wine variety. It ripens relatively early, usually after Rondo and before Pinot noir and generally yields quite well; 7.5–10 tonnes-ha (3–4 tonnes-acre) are achievable on a good site. For deep red wines, it has a higher quality than wines made from Dornfelder or Rondo, although it probably best used in a blend of all three. It takes oak well, has great colour, soft tannins and its acids are manageable. It gained very high scores in the PIWI International Wine Competitions. Its slow expansion in GB suggests that other varieties for red wine, Pinot noir and Blauer Frühburgunder (Pinot noir Précoce), plus the red PIWIs (Cabernet Cortis, Cabaret noir, and Divico), which have even better natural resistance, are taking its place. Probably not a first-choice variety to plant today. Growers interested in its wine quality should be able to find wines from both GB and other cool regions so that they can make a personal assessment of its style and quality.

58 The Regent diamond was in the same gallery in the Louvre that was raided in 2025, but the thieves either missed it, or considered it too problematic to steal.

Réselle

Type: PIWI
Colour: White
Origin: Blattner, Valentin
Crossing: Bacchus x Seyval blanc
Date of crossing: Not known
Year of registration: Not known
Planted area in Germany (2023): None
Planted area in GB (2024): None

Réselle is another Blattner crossing, this time of Bacchus x Seyval blanc (the same parents as Calardis Musqué) and one which Blattner himself rates very highly. He has 1-ha of the variety in his own vineyards in Soyhières, Switzerland. It is resistant to both mildews and wine quality, with zesty citrus and lime notes, can be very high. The variety is sometimes called Recele or VB 86-3. It would be well worth experimenting with in GB.

Riesel

Type: PIWI
Colour: White
Origin: Blattner, Valentin
Crossing: Riesling x Unknown resistance partners
Date of crossing: 1989
Year of registration: Not known
Planted area in Germany (2023): None
Planted area in GB (2024): None

Another Blattner crossing, this time based upon Riesling with unknown resistance partners. The bunch is compact without shoulders and the grapes turn deep gold when ripe i.e. very similar to Riesling. Growth is strong and upright, and deleafing in the fruiting zone is advised. It is said to be frost hardy (Freytag). It is early ripening with low acidity and high sugar levels are possible, making it ideal for cooler climates such as GB's. It requires a basic spray programme. Wine quality is high. Very little is known about this variety, but on the basis of its pedigree it would be worth a trial plot. According to the *Vitis International Vine Variety* database, the crossing is with Riesling, not Cabernet Sauvignon as stated in several other sources.

Rinót

Type: PIWI
Colour: White
Origin: Mendel University, Brno, Czech Republic
Crossing: Merzling x (Villard blanc x Pinot gris)
Date of crossing: Not known
Year of registration: 2008
Planted area in Germany (2023): None
Area planted in GB (2024): None

A complex crossing from Dr Miloš Michlovský at the Mendel University, Brno in the Czech Republic (formerly Czechoslovakia). Merzling is a cross of Seyval blanc, Riesling and Pinot gris on the one side, plus more French hybrids and Pinot gris on the other. It is also related to Solaris. It has excellent natural resistance against both Powdery Mildew and *botrytis* and good resistance against Downy. The bunches are loose with medium sized grapes which allow for natural ventilation, thus helping them keep healthy. With this degree of natural protection the grapes can hang for a long time, thus allowing sugars to rise and acids to soften. The wine is lightly fruity with balanced acidity and a good length. Volker Freytag, the nurseryman that grafts more PIWIs than any other German nursery and has worked closely with Blattner over thirty years, rates it highly and Rinót was chosen as one of seven of his 'Pioneer Wines'. He suggests it would be ideal for GB, especially for sparkling. I agree.

Rondo

Type: PIWI
Colour: Red
Origin: Geisenheim (via Lednice Viticultural College)
Crossing: Zarya Severa (Severnyi) x St Laurent
Date of crossing: 1964
Year of registration: 1997
Planted area in Germany (2024): 7-ha
Planted area in GB (2024): 64.23-ha

Of all the varieties with complicated histories attached, Rondo must have one of the most complex. The original *Vitis Amurensis* vines, from which Rondo is partly derived, were wild vines that came from Manchuria, in the north of China, where the River Amur marks the border between China and Russia. Here, on account of the early onset and severity of the winters, wild vines need be able to withstand deep winter temperatures and therefore colour up and ripen early. In 1910, a Russian vine breeder, Ivan Vladimirovich Michurin (1855–1935) – pictured below – selected an open-pollinated seedling of Précoce de Malingre, which was probably pollinated by a *Vitis Amurensis* in Michurin's vine collection, which he called Seyanets Malengra.

Michurin was a famous plant breeder whose work was recognised by Lenin and whose private research station at Tambov, about 435-km south-east of Moscow, became the Michurin Central Genetic Laboratory in 1934. Seyanets Malengra is a female vine and is used in Russia and Ukraine as a table grape variety. In 1936, two Russian plant breeders, Yakov Potapenko and E. Zakharova, working at the All-Russia Research Institute of Viticulture and Winemaking (later called the Potapenko All-Russia Research Institute of Viticulture and Winemaking), situated at Novocherkassk in Rostov Province, crossed Seyanets Malengra with another *Amurensis* vine to produce a variety called Zarya Severa (also sometimes known as Severnyi). This crossing then found its way to Lednice in Moravia, Czechoslovakia where Professor Dr Vilém Kraus crossed it, in 1964, with St Laurent, an old Austrian variety. This crossing then travelled to Geisenheim where Professor Helmut Becker improved it through clonal selection and gave it a breeding number Gm 6494/5. With me so far? It was then trialled at Geisenheim and proved capable of producing good crops of early ripening, deeply coloured grapes. However, it was never widely planted in Germany and in 2019 there were only 11-ha and in 2024 there are still only 11-ha.

In 1983 Professor Becker gave me 50 vines to plant at Tenterden, as well as 50 each to Ken Barlow at Adgestone and Karl-Heinz Johner at Lamberhurst, to see what the results would be. From the first harvests it showed itself to be well adapted to our conditions, and plantings have been increasing since then. From 2-ha in 1990, it has continued to grow in popularity and stands at around 64.23-ha in 2024. Ignoring Pinot noir and Meunier (from which little red wine is made) Rondo is the second most widely planted red variety in Britain. In other cool climates such as Belgium, Denmark, the Netherlands, and Sweden it is one of the most important varieties. Ireland's largest vineyard, the 0.90-ha Thomas Walk Vineyard near Cork, first planted in the 1980s, is 100 per cent planted with Rondo, which they call '*Amurensis* Walk'. For the first decade of its life in Britain it was called Gm 6494/5, but eventually the EU gave it a name – Rondo – and we persuaded DEFRA to classify it as a *vinifera* (as the Germans had done five years earlier).

Initially, Rondo had all the makings of a good variety for Britain: early to ripen, with a good crop of deep red grapes with low(ish) acidity and wine quality that was certainly better than anything else seen in the mid-1980s – Léon Millot, Maréchal Foch and Triomphe d'Alsace (as it was then called)

- being the alternatives. Chapel Down made two versions: a straightforward wine and a Reserve, both from grapes from Anthony Pilcher's Chapel Farm vineyard on the Isle of Wight. The Reserve, with enough oak and some time, could be very good, although the flavour was always very slightly unusual. I always likened it to a Syrah-Tempranillo blend - the American oak helped - and I tried to persuade myself for years that it had a future. However, reluctantly, and in the face of the onslaught from Blauer Frühburgunder (Pinot noir Précoce), Pinot noir and Regent, I do not think the wine quality is high enough for it to succeed in the long term.

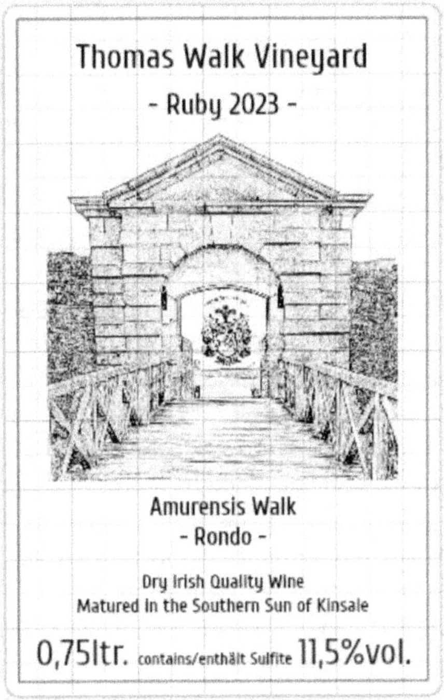

It is a very vigorous variety, with big leaves (almost like the leaves of kiwi fruit) and canes that can stretch for several metres if allowed to run. Consequently, canopies tend to be crowded and shaded and although a PIWI, it is still fairly susceptible to *botrytis*. The wine colour is excellent, deeper than Dornfelder and Regent and the quality is passable, but not of the first order and certainly not what the consumer expects of a wine of that appearance. Although the area planted has expanded, most of this additional planting is in vineyards in the more challenging parts of Britain and I do not believe that any mainstream vineyards view it as a serious variety for the future. Although I value the direct link back to Helmut Becker and those first 50 vines, sentimentality never helped make good wine and plantings of Rondo, although they will probably expand, will do so only slowly. Growers interested in its wine quality should be able to find wines from both GB and other cool regions so that they can make a personal assessment of its style and quality.

Sauvignac

Type: PIWI
Colour: White
Origin: Blattner, Valentin
Crossing: (Sauvignon blanc x Riesling) x Unknown resistance partners
Date of crossing: 1991
Year of registration: 2018
Planted area in Belgium (2023): 13-ha
Area planted in GB (2024): 6.04-ha

Sauvignac is one of the more widely grown of the Blattner varieties and stems from the later 'Cal' series of crossings, almost all of which are red. It has the breeding number Cal 6-04. These crossings have greater disease resistance than some of the earlier crosses and will give good control over both mildews and *botrytis*. The grapes are quite small (like Riesling) with thick skins, and the bunch is quite loose. The variety is one of Volker Freytag's seven 'Pioneer Wines'. Wine quality is high with good *vinifera* characteristics. This is definitely a variety for GB and as shown by the area already planted, it is showing good distribution. It is only planted on 14 separate vineyard parcels in GB (Essex having the most parcels) and nearly all of the others are in the west and south-west. Definitely a variety to watch.

Sauvignon Rytos

Type: PIWI
Colour: White
Origin: University of Udine and Institute of Applied Genetics (IGA)
Crossing: Sauvignon blanc x Bianca
Date of crossing: 2002
Year of registration: 2014
Planted area in Germany (2023): None
Area planted in GB (2024): None

Sauvignon Rytos is a relatively new variety which Vivai Cooperativi Rauscedo (VCR) have in their stable of PIWIs. One of its parents is Sauvignon blanc which gives the variety its wine quality and style. The other parent is Bianca, a Hungarian cross which includes Villard blanc which gives it its 'good to high' resistance to the mildews. It has a tight cluster, which means that spraying against *botrytis* is required around flowering time, especially before bunch close. It should ripen at the same time as Pinot noir. Acids are moderate and sugars can be high. Wine quality is good to very good, hence the recommendation.

Seyval blanc

Type: Hybrid
Colour: White
Origin: Seyve-Villard Nursery, St Vallier, France
Crossing: Seibel 5656 x Seibel 4986
Date of crossing: 1921
Year of registration: Not known
Planted area in GB (2024): 122.71

Seyval blanc, a crossing made by Bertille Seyve (the younger) at the nursery of his father-in-law, Victor Villard, in St Vallier on the Rhône in 1921, was originally known as Seyve-Villard 5276. In some references, the crossing is given as Seibel 4996 x Seibel 4986, but Pierre Galet (who was probably the most famous ampelographer in the world) assured me that this was incorrect. He told me in a letter he wrote me that Bertille Seyve's son had stated in the *Viticulture Nouvelle* of 1961 (p. 182) that the crossing was Seibel 5656 x Seibel 4986 (S. 4986 is known as Rayon d'Or). In 1958 there were 1,309-ha in France, and it appeared to be declining. However, Anderson in his 2020 revision of *What wine-grape varieties are grown where*[59] states that worldwide there were 2,699-ha, with 2,584 in France, and most of the balance in Canada and GB. Its is also grown in New York State and Ontario, plus a small amount in Switzerland. Surprisingly, there appears to be very little grown in the Scandinavian or Baltic countries, and I am not sure quite why. In GB it has shown a steady rise from a low point of 88-ha in 2015, to today's heady total of 123-ha. No doubt as more and more growers get to like it and get to like its disease resistance qualities and yields, it will continue to expand. It is currently Britain's fifth most widely planted variety but will probably be overtaken by Solaris quite soon.

It was first grown in Britain by Edward Hyams who, in 1947, obtained vines directly from the Seyve-Villard nurseries (as they had become known by then) and planted them in his small vineyard at Molash, near Canterbury, Kent. Brock also imported the variety from the Swiss Federal

59 *What Wine Grape Varieties are Grown Where* by Kym Anderson and Signe Nelgen, University of Adelaide, 2020.

Vine Testing Station at Caudoz-sur-Pully a year later, in 1948. These two pioneers tested it and soon found that it was a very suitable variety for our climate and. together with MT, it became the standard variety for almost all the early vineyards. When Sir Guy Salisbury-Jones planted his vineyard at Hambledon in 1952 he chose Seyval blanc on the advice of a helpful Burgundian grower whom he sat next to at a *Confrérie des Chevaliers du Tastevin* dinner.

Seyval blanc has many attributes. It sets good crops even in cool years, is not vigorous and has a good open habit with small leaves and only a few side-shoots. It is unusual in that the flowers appear and are very prominent before many of the leaves really develop. Its disease resistance is good, although with climate change and the rise in natural sugar levels, it has become a bit more prone to *botrytis*. It ripens later than Pinot noir, but usually before Chardonnay. Last, but by no means least, it yields well – in some years very well – and 20 t-ha (8.0 t-acre) and more is not unknown. The grapes used to be pretty low in natural sugar and I remember plenty of years in the 1970s and 1980s when only 5–6 per cent natural alcohol was reached. However, global warming has increased sugar levels and while not as high as Bacchus, are more than adequate, especially for sparkling base wine. The grassiness that used to typify some Seyval blanc wines seems to have disappeared, and growers find it a useful variety for blending with a range of varieties. It also takes to oak-ageing well, is a good foil for lees-ageing and *battonage* and is great for sparkling wines. It can suffer from *botrytis* towards the end of ripening, so canopies need to be open and it is probably wise to spray at least twice before bunch close.

Seyval blanc suffers from a bad press and – rather like Meunier in Champagne – it's a useful variety, but do not let's frighten the horses by talking about it too much. Most wine writers and many wine merchants, faced with the word 'hybrid', seem to recoil into their shells a bit, strike an arrogant pose, and declare without a *shred* of evidence, let alone experience in winegrowing or winemaking, that 'of course you cannot make good wine from a *hybrid*'. The truth is, of course, that you can and people do. Wines made from Seyval win their share of medals and awards: I won the Gore-Browne Trophy with Seyval twice (in 1981 and 1991) and Colin Gillespie at Wootton won it in 1986 with a Seyval blanc. It can still be found in some good sparkling wines (Breaky Bottom, Camel Valley, Denbies, and Stanlake Park. for instance) as well as a lot of good still wine.

Breaky Bottom vineyard in East Sussex, home to some of the best Seyval blanc sparkling wines

When I imported my first Seyval, I obtained stock from one of Professor Becker's friends, Ernest Pfrimmer who had a nursery at Hurtigheim, near Strasbourg, and the results were always good. When I started importing a lot of Seyval vines from France, I spent a considerable time searching for a good source-block of vines and together with François Morisson (from the Morisson-Couderc Nursery) found an ideal one in Blois on the Loire. From this vineyard he took wood and grafted many thousands of vines for me. Having seen a fair few Seyval vineyards, both in Britain and overseas, I am convinced that there is quite a bit of clonal variation and some of Britain's vineyards are planted with inferior clones. This might account for some of the grassy, herbaceous notes that are sometimes encountered although that could just be harsh pressing. I have seen some Seyval in a GB vineyard that is plainly another variety altogether

– possibly another Seyve-Villard hybrid – and clearly unsuited for British conditions. In 2013, I travelled to Switzerland to visit Blattner and saw an excellent mother-block of Seyval, and I now only import vines produced with scion wood originating from this vineyard.

The slight revival in Seyval's fortunes give me some encouragement that its future is assured, albeit as a minor but useful variety. The late and much-missed Peter Hall of Breaky Bottom vineyard probably did more than anyone to raise the profile of Seyval blanc-based sparkling wines and my wish is that his legacy is that the variety should continue to prove its worth in years (and decades) to come. It is especially suitable for organic and biodynamic producers, plus those in the climatically challenged parts of GB.

Solaris

Type: PIWI
Colour: White
Origin: Freiburg
Crossing: Merzling x Gm 6493
Date of crossing: 1975
Year of registration: 2001
Planted area in Germany (2024): 230-ha
Area planted in GB (2024): 117.07-ha

Solaris is a Freiburg interspecific cross, once known as Fr. 240-75, bred by Dr Norbert Becker in 1975 and released in 2001. Its parents are Merzling and Gm 6493. Merzling is a crossing between our old friend Seyval blanc and a (Riesling x Pinot gris) crossing. Gm 6493 is a Geisenheim *Vitis Amurensis*-influenced crossing and is Zarya Severa x Muscat Ottonel, Zarya Severa being one of the parents of Rondo! In 2012 there were 194-ha planted in Germany (mainly in Baden, the Pfalz and the Mosel), and by 2024 this grew to 230-ha, making it their fourth most widely planted PIWI. It is also planted in the Trentino-Alto Adige region of Italy, and in Switzerland and is a major variety in the cooler countries of Europe such as Belgium, the Netherlands and the Scandinavian countries. It is by far the most widely planted white variety in Denmark. In Ireland it, and Rondo, are the only two varieties to give anything like viable crops.

The *Taschenbuch der Rebsorten* says that it is very vigorous, with very large leaves, and that it ripens early with very high sugars, usually over 100°OE (13.8 per cent potential alcohol) and often nearer 130°OE (18.4 per cent potential alcohol). It also states that the acidity is less than Ruländer (Pinot gris) – which would mean that its acidity is towards the bottom of the acid spectrum, i.e. 5–6 g/l as tartaric acid. It has very good resistance against both mildews, and black rot and is also especially resistant to *botrytis* to which high-sugared varieties are usually prone, and its loose bunches help in this respect. It sounds very similar in many ways to Reichensteiner, but with better disease resistance. Wasps can be a problem (as with any early ripening variety) and good wasp nest control needs to be practised if real damage is to be avoided.

In Britain it is planted today (2024) on 117.07-ha which is up from 16-ha in 2013, a greater rate of expansion than any other variety, and it is the most widely planted PIWI. It will surely catch up and overtake Seyval blanc very soon. There are plantings spread across the country, and WS lists 215 separate vineyard 'parcels'. The majority are in East Anglia, the West and Midlands & North regions, but there are a few larger plots in Kent and West Sussex. Wales has 13-ha of it.

In terms of wine quality, it has done well in competitions with at least one gold, several silvers and many bronzes in the WineGB annual competition. It gained the highest scores in the PIWI International Wine Competition results for the years 2022-2024. As vineyards expand ever further north – productive Scottish vineyards cannot be far away – I see Solaris being planted more widely still. Another minor, but important variety, especially for those wishing to reduce the number of sprays they have to apply. Growers interested in its wine quality should be able to find wines from both GB and other cool regions made from Solaris so that they can make a personal assessment of its style and quality.

Soreli

Type: PIWI
Colour: White
Origin: University of Udine and Institute of Applied Genetics (IGA)
Crossing: Tocai Friulano x Kozma 20-3
Date of crossing: 2002
Year of registration: 2010
Planted area in Germany (2023): None
Planted area in GB (2024): 0.30-ha

Soreli is a relatively new variety which Vivai Cooperativi Rauscedo (VCR) have in their stable of PIWIs. Kozma 20-3 is a Hungarian crossing of Bianca (a complex muscat crossing) with another Hungarian cross with both muscat and *Amurensis* parentage. It has 2 genes for resistance to both Powdery Mildew and one gene against Downy. It also has good winter hardiness. The *Amurensis* parentage accounts for its early ripening and low acidity, all good for GB conditions. Its growth habit is said to be 'semi-upright or drooping' so will need good canopy management and timely leaf removal. Its inclusion in my recommended list is based on its very good wine quality. It is a sibling to the variety called Fleurtai. Good for both still and sparkling.

Souvignier gris

Type: PIWI
Colour: White
Origin: Freiburg
Crossing: Seyval blanc x Zähringer
Date of crossing: 1983
Year of registration: 2012
Planted area in Germany (2024): 566-ha
Planted area in GB (2024): 4.61

Souvignier gris is Seyval blanc crossed with a variety called Zähringer which is a cross of Gewürztraminer and Riesling. It gets all its disease resistance from Seyval and its wine qualities from the two high-quality *viniferas.* At one time it was thought to be a Cabernet Sauvignon x Bronner variety, but DNA analysis has discovered otherwise. It has excellent resistance against both mildews (better against Downy than Powdery) and *botrytis* but does need a minimum spray programme. It has a loose bunch and the grapes turn dark pink as it ripens.

It is certainly one of the most successful of the PIWIs and apart from Germany with 566-ha, it is planted in Austria (76-ha), Switzerland (45-ha), Belgium (33-ha) and even France (388-ha). It is also planted in almost all the other cooler regions in Europe. The reasons? Its wine quality and its yield in tough conditions. It has better than average wine quality, quite like a Chardonnay-Muscat blend and it received the third highest scores in the PIWI International Wine Competitions, closely behind Muscaris and Regent. I feel it has a good future in GB for the specialist grower. I also like the name.

Voltis

Type: PIWI
Colour: White
Origin: INRA Bordeaux
Crossing: Villaris x VRH 3159-2-12
Date of crossing: 2002
Year of registration: 2018
Planted area in France (2023): 5.75-ha
Planted area in GB (2024): 1.02-ha

Voltis is one of a small selection of new PIWI varieties produced in France and now being grown there and it is the first new variety to be allowed to be grown in the Champagne region since the *appellation* was formed. In Champagne it can occupy up to 5 per cent of the vineyard area (I assume per grower) and can be blended with other varieties to make up to 10 per cent by volume of the *cuvée.* A final decision on the variety's suitability for Champagne will be taken in 2033, after which the variety will be accepted, or banned (and if banned, all vines will have to be removed). It is said to ripen at the same time as Chardonnay although at least one report said it ripens later. It is said to have 'very strong resistance' to the major diseases. It is currently only grown on 5.75-ha in France, and the first harvests have only just been picked.

The reason the Champenoise need a new variety is twofold: to cut back on the number of sprays they need to apply; and secondly to have a variety that hangs on to its acidity as it ripens as Chardonnay in warm years tends to lose it quite quickly. Time will tell whether it gets any traction in its home country. Extensive trials with the variety have already taken place and the results (as a blending partner) are said to be positive.

As for recommending it for GB, I hesitate to say 'plant tens of hectares' but it would certainly be worth a trial if you had a warm site. Not advised for cooler GB regions as acids are high.

Chapter 9
Rootstocks for Great Britain

Note: Vines can be grown on their own roots but they are then susceptible to infestation by *Phylloxera* and it is therefore not recommended. There is more on this in Chapters 16 and 17 of *Wine Growing in Great Britain.*

Young vines in their nursery beds in the Ardèche region in France

The question of which of the many dozens of rootstocks used around the world are best for vineyards in Britain is an interesting one. Like the choice of varieties, this task is made harder by the fact that no rootstock has been developed for British conditions and that all rootstock recommendations are based upon data derived from climates which are both warmer and drier than Britain's, from regions where yields are typically higher (often much higher) and more regular, and in soils which may differ considerably from ours. The behaviour of a rootstock in a lean, dry soil, where there is little summer rainfall, where temperatures are often over 30°C and where yields are both higher and more consistent than in Britain, will not be the same as in a deep, fertile soil, well-watered with summer rainfall and where yields might be modest one year, negligible the next and well above average the year after. Likewise, the behaviour of a rootstock in a region where the climate is hot, vines are irrigated and yields are consistently high. Therefore, whenever looking at recommendations for rootstocks to use in Britain, these differences in climate and growing conditions need to be taken in to account.

Rootstock tables (such as the one at the end of this chapter) will usually give you several recommendations concerning not only their active calcium carbonate ($CaCO_3$) tolerance and root growth habit, but also their 'vigour' and 'influence upon maturity'. As I have said above, these recommendations are not based upon British data (another job for Plumpton perhaps?) and therefore a 'health warning' needs to be issued before relying too much upon the tables you can find on the internet and issued by overseas vine nurseries. In my experience, there are other factors which can also have an influence upon the behaviour of the vine, not only the rootstock. The weather during the season undoubtedly has a huge impact upon the vigour of vines: a cool, wet season will promote growth; a hot, dry season will almost certainly curtail vigour. The other very important factor is yield. A vine bearing 5 tonnes-ha (2 tonnes-acre) is going to be more vigorous than one carrying 10–15 tonnes-ha (4–6 tonnes-acre), especially if this is coupled with a warm, dry year (i.e. the sort of year likely to lead to high yields). In 2013 British vineyards were subject to very unusual circumstances. We had a very late spring, an ideal flowering period which gave high yields, a warm spell in the middle of the summer, followed by a cooler, wetter, less sunny October and the latest harvest ever experienced. How could you expect a rootstock to conform to a description based upon completely different circumstances?

In the early days of modern British viticulture, with the exception of a few growers who obtained their vines from France, the bulk of vines planted up until the early 1990s were German varieties, of German origin and consequently, on German rootstocks. The four standard German rootstocks are 125AA, 5BB, 5C and SO4, with Teleki 8B and Börner bringing up the rear. All of these are classified as 'high' to 'medium-high' in terms of vigour and, coupled with already vigorous varieties such as Bacchus, Huxelrebe, Kerner, Müller-Thurgau, Ortega, Reichensteiner and Schönburger, tended to produce, in our cool climate and our fresh, loam-rich, well-watered soils, some exceptionally large, leaf-endowed canopies which did little to help the problems associated with shading and over-vigorous canopies. Of the 'old' varieties, only Seyval blanc, being a naturally low-vigour variety, did not suffer from over-vigorous canopies. This is one of the reasons why, of almost all of the varieties popular in the early days, Seyval has maintained its position high up in the list of varieties being grown in Britain. Its above-average yield has also helped its survival.

The arrival in Britain of the Champagne varieties has somewhat altered the rootstock situation and, in my view, made a significant impact upon the vigour of our canopies. In France, the origin of the majority (but by no means all) of the Chardonnay, Pinot noir and Meunier vines now being grown in Britain, the most popular rootstocks for these varieties are SO4, Fercal and 41B in high active CaCO3 soils (with 161-49 as another option) and 101-14, 3309C and 420A in soils with 20 per cent or less active CaCO3. With the exception of SO4 and Fercal (both classified as having 'medium' vigour) all of the others are classified as 'low' or 'low medium' in terms of vigour. On paper, this may seem a very marginal difference, but it is my experience that these marginal differences, in the vineyard, make an appreciable difference to vigour of shoot growth and therefore to the density of the canopy. It worries me sometimes that growers prefer to get their Chardonnay and Pinot noir from Germany and on German rootstocks without realising the difference the rootstock can make.

The first thing you need to know before you can select a rootstock is whether you are likely to need lime-tolerant rootstocks. In soils of pH 7 or lower, and without any obvious signs of chalk or limestone, the level of active CaCO3 in your soil is likely to be low. For soils of over pH 7, the best and safest advice is to test for the level of active CaCO3. This is NOT the same as the level of total CaCO3 – they are related, but not in a linear fashion and it is not, in my view, safe to rely on taking 50 per cent of the total level of CaCO3 as being the level of the active CaCO3 (something I have seen suggested is acceptable) as this is not accurate enough. The reason that the level of active CaCO3 is important is that if the rootstock selected does not have the right level of tolerance, the vines will get chlorosis, a condition where they are unable to access iron in the soil and are therefore unable to produce chlorophyll, necessary for photosynthesis to take place. Vines suffering from chlorosis will turn light green, then yellow, and then almost white, before starting to die. Vines suffering from chlorosis can be treated by spraying or soil drenching with chelated iron, iron that has been treated so that it is soluble (Ferleaf and Fersoil are the two best known products), and while this may be a temporary solution, it is much better to select the right rootstock in the first place. In France, rootstocks are given an 'IPC' rating (*l'Indice de Pouvoir Clorosant*) which runs from 0 to 120, with rootstocks at the lower end (0–20) being safe in soils with no or low levels of active CaCO3 and those with higher ratings (30–120) being better in soils with high levels of active CaCO3. I have never used this index myself and in practice find that the active CaCO3 measurement is more than adequate.

When it comes to active CaCO3 levels, I would prefer to err on the side of caution as, however well done, sampling a field can only tell you about a small section of the whole field and there may well be patches of ground where levels are higher than that shown by sampling. Therefore, for levels of active CaCO3 of 15 per cent of more, I would only plant on 41B or Fercal (or maybe 161-49). 41B to me is the better rootstock, even though it is listed as having a delaying effect upon maturity. I find it less vigorous than Fercal. Unless you are really sure about the accuracy of your soil sampling, I would avoid SO4 at active $CaCO_3$ levels of 15 per cent or more, even though the tolerance of SO4 is generally stated to be 20 per cent. Rootstock 161-49 is also an option, but while in some situations I have found this to be excellent, producing a low-vigour canopy and good crops, in other situations it has struggled, especially when paired with Chardonnay. With Pinot noir and Meunier it has proved to be better. Rootstock 161-49 suffers from what is known as tylosis (and which the French call *thyllose* or sometimes *folletage*) which can result in a higher percentage of early vine deaths than would normally be expected with other rootstocks. I would therefore advise caution with 161-49, especially in situations where the vine might be put under stress in its first two years. Such situations might be poorly drained sites, adverse soil conditions, drought, weed competition, or early cropping. If 161-49 is used, I would recommend keeping the young vines as weed-free as possible and cultivating the inter-row areas for the first two full growing seasons before grassing down.

In situations where levels of active $CaCO_3$ are lower than 15 per cent, my first choice would be 101-14, 3309C and 420A, all of which are at the lower end of the spectrum for vigour and have a slight advancing effect upon maturity. If these were not available, I would select 5C, 125AA and SO4, listed in increasing order of vigour, with SO4 being the most vigorous. However, in soils with low levels of active $CaCO_3$, there is nothing stopping your using 41B or Fercal and I have planted plenty of vines in soils with pH 6.5 using these rootstocks. For very acid soils i.e. pH 5.5 or below, the only rootstock that appears to be recommended in the literature is Gravesac. In some sources it is suggested that 101-14, 3309C and SO4 might also be suitable, although in other sources this is contradicted, and before one were to embark on a large-scale planting on soils with this level of pH, some further research on rootstocks would be advised. Luckily there are few soils in British fruit-growing regions with this level of pH and, in any event, soils with pH levels below 6.0–6.5 must be limed in an attempt to bring them up towards neutral (pH 7).

Two other factors must be considered when selecting rootstocks: soil and vine variety. With deep, free-rooting soils, the best rootstocks are going to be ones with lower vigour. This does not mean that more vigorous rootstocks cannot be used, but in order to have an open and airy canopy, you will probably need to do more canopy management (trimming and deleafing) than with lower-vigour rootstocks. Likewise, with vigorous varieties such as Bacchus, Reichensteiner, Müller-Thurgau and other German varieties, I would always aim for lower-vigour rootstocks, although 101-14, 3309C and 420A are seldom available from German nurseries unless custom-grafted.[60] Pinot noir can also be quite vigorous and is best on a lower-vigour rootstock, although Chardonnay appears to do well on both. Only with Seyval blanc, which is naturally low in vigour, might I look for a rootstock with a slightly higher vigour potential, and I have known it do well on SO4 and 5BB. On deep, vigorous soils, 420A also suits it.

Apart from those mentioned above, there are many other rootstocks which may be perfectly acceptable in British vineyards. I have used both Gravesac and Riparia Gloire de Montpelier (RGM) in vineyards and they performed well, although RGM does not appear to be that low in vigour in our climate. However, these two rootstocks are not widely used and are not always easy to obtain. Yalumba Nursery has both an excellent 'Rootstock Selector' as well as a lot of information about individual rootstocks on its website (www.yalumbanursery.com), although of course aimed at Australian growers. Mercier Nursery also has good information in English on their Californian website (www.mercier-california.com) under

60 Custom-grafting is where you ask a nursery to graft a rootstock and scion variety which is not one they usually produce. They usually want a minimum order of at least 2,000 vines and the outcome is almost always either higher or lower than their estimate.

'Varieties' and then 'Rootstocks'.

A final warning concerns vine nurseries. They are in the business of selling as many vines as possible from the vines they graft and seeing those vines survive and thrive in their client's vineyards. Some rootstocks graft more easily than others, giving a higher percentage of 'take', meaning that once lifted and on the sorting table, far fewer are rejected. Rootstocks with higher vigour also tend to root more easily, giving better roots at lifting from their seven to eight months in the nursery and also tend to put down roots more easily once planted out in the vineyard. Fercal and SO4 are well known as falling into this category and are well liked by nurseries. However, that does not mean that these vines always produce the best vineyards and the best crops in the long term. Under British conditions, with most situations and vine varieties, I prefer a rootstock with a slightly lower-vigour potential.

Rootstocks for vineyards in Great Britain

Rootstock	Crossing	Active $CaCO_3$ tolerance	Root growth	Vigour	Influence on Maturity
Riparia Gloire (RGM)	Riparia	6%	Shallow	Low	Advances
101-14	Riparia x Rupestris	10%	Shallow	Low	Advances
3309C	Riparia x Rupestris	11%	Semi-deep	Low-Medium	Advances
Gravesac	161-49 x 3309C	15%	Semi-deep	Medium	Advances
Börner	Riparia x Cinerea	15%	Semi-deep	Medium-High	No effect
125AA	Berlandieri x Riparia	17%	Deep	Medium-High	No effect
5C	Berlandieri x Riparia	17%	Deep	Medium-High	Slightly Advances
420A	Berlandieri x Riparia	20%	Semi-deep	Low-Medium	Slightly Advances
SO4	Berlandieri x Riparia	20%	Semi-deep	Medium+	No effect
Binova	SO4 Mutation	20%	Semi-deep	Medium++	No effect
161-49C	Riparia x Berlandieri	25%	Semi-deep	Low-Medium	Advances
41B	Chasselas x Berlandieri	40%	Shallow	Low-Medium	Delays
Fercal	Berlandieri x Riparia	45%	Deep	Medium	Advances

Note: Recomendations on root growth, vigour and influence on maturity are based upon non-UK data

Table 37

Appendix 1

The Complete Story of Sparkling Wine production in Great Britain 1662-2024

The production of sparkling wine in Britain – although not from home-grown grapes – is verifiably more than 350 years old, and we know from the two papers read at the newly-founded Royal Society in December 1662 that sugar added to a fermented product and sealed in a bottle with a tightly bound stopper produced a 'brisk and sparkling' product.[61] The Reverend John Beale's 'Aphorisms on Cider', read to the Royal Society on 10 December 1662 says that 'bottling is the next best improver' for cider and that 'two to three raisins into every bottle' plus 'a walnut of sugar' – a recipe guaranteed to produce a secondary fermentation – works wonders on the cider. A week later, on 17 December, it was the turn of the now famous Dr Christopher Merrett to read his paper, 'Some Observations Concerning the Ordering of Wines' and describe how Britain's seventeenth-century 'wine coopers' were making their wines 'brisk and sparkling' by the addition of sugar. This practice was certainly happening before 1662 and followed the development of the strong *verre Anglais* bottles which Sir Kenelm Digby had (probably) been involved in perfecting since around 1628-1630[62].

Exactly when the first sparkling wine made from English grapes was produced is open to debate.[63] Certainly, wines being made in England in the 1750s were considered comparable to Champagne and as has been documented by me in other books, the wines produced at Painshill Place between 1741 and 1779 were often described as such. Of course, Champagne in those days was not always the sparkling wine that we know today. I have a wine list from the *Magaszin de Vins Fins Chez Terral* from Pontac, a village just outside Bordeaux, dated 1760, which lists *Champagne mousseux* and *Champagne non-mousseux* both at the same price.

The first recorded production of bottle-fermented sparkling wines made from British-grown grapes is probably that carried out by Raymond Barrington Brock at his Oxted Viticultural Research Station in the 1950s. The *Daily Mirror* of 17 August 1950 carried an article entitled 'A bottle of Maidstone '49' which praised the work of Brock and that other viticultural pioneer, Edward Hyams and ended by saying, 'perhaps ten years hence you'll be raising a glass of sparkling Canterbury [Hyams lived in Shottenden, near Canterbury and had a small vineyard there] in honour of the men who made an English wine industry possible.' In September 1959 Brock welcomed members of the wine trade to his vineyard for a tasting and offered them a number of different wines, including sparkling wines. I have a letter dated 11 September 1959 sent to Brock from John Clevely, then a young Master of Wine, in which he thanks Brock for the visit and tasting and ends with a postscript saying:

61 This is believed to be the first use of the word sparkling in English in relation to wine or cider.

62 My book *The Knight Who Invented Champagne* is all about Digby and his bottle.

63 A sparkling wine called *Moussec,* originally made from grape concentrate produced from grapes grown in the Aube region of France, was produced initially in a factory in Rickmansworth, Hertfordshire from the mid-1930s until well into the 1980s. The original label said it was made from 'Champagne grapes' and produced by the 'Champagne method', but these claims were later dropped. Given that it was made in multiple bottle sizes, including single-serve bottles, it was probably made in the tank method (if not also carbonated). The company was founded by a Monsieur Rivollier from the Aube region who made the concentrate from grapes grown in his region but fell out with the authorities in Champagne and was forced to stop. The company was bought by Reckitt and Coleman (R&C Vintners) who continued production in their winery in Norwich on Carrow Road. For many years the alcohol content was labelled as being '21° Proof', a term usually used for spirits at the time. This level of 'proof' is around 10.5% abv.

'Moët must look to their laurels if you really start going "commercial" with that sparkling wine. I thought it was wonderful.' Praise indeed. Some of these sparkling wines survived undisturbed in the Station's cellars until the 1980s.

Sir Guy Salisbury-Jones at Hambledon, whose initial (1953) plantings included 20 Chardonnay vines, experimented with the production of a bottle-fermented sparkling wine, and in 1969 Bill Carcary, his vineyard manager, produced a batch of 60 bottles. Salisbury-Jones expanded the plantings of Chardonnay in 1970 with a further 1,000 vines but whether to make still or sparkling wine is not known. In 1979 his winemaking consultant Anton Massel helped produce a batch with apparently favourable results and as Salisbury-Jones also grew Auxerrois and Meunier, which ripened more easily, these became the basis of their sparkling wine *cuvée*. However, Sir Guy considered that the production costs were too high and the length of time the wine needed to mature was too long to make the product commercially viable and production ceased. Little did Bill Carcary know it at the time, but his experiments were to be the foundation stone of Britain's highest profile sparkling wine producer, Nyetimber.

The first producers to make commercial quantities of bottle-fermented sparkling wines were Nigel (de Marsac) Godden at Pilton Manor Vineyard in Somerset – first planted in 1966 – and Graham Barrett at Felsted (or Felstar[64] as it was originally known) Vineyard in Essex – first planted in 1967. As was quite usual at that time, the main varieties grown were Müller-Thurgau and Seyval blanc and it is probable that it was these varieties that were used. Their wines – never produced in large volumes – were certainly interesting, maybe even worth drinking and in the 1979 English Wine of the Year Competition (EWOTYC) the *1976 Felstar Méthode Champenoise* won a silver medal and the following year, 1978, the *NV Pilton Manor De Marsac Brut Méthode Champenoise* won a bronze. These early successes, however, didn't seem to help sales much and the production of both faded out.

1983 Carr Taylor starts making bottle-fermented sparkling wine

The next appearance of a bottle-fermented sparkling wine in the EWOTYC (ignoring the carbonated *1983 Barton Manor Sparkling Rosé* that won a gold medal in the 1984 competition – delicious though it was) was in 1987, when the first Carr Taylor sparkling wine won a medal. David and Linda Carr Taylor first planted vines at their vineyard in Westfield, near Hastings, East Sussex, in 1973 and until 1983 their grapes were sent to Lamberhurst Vineyards for winemaking. From their huge 1983 vintage however, when Reichensteiner cropped at 15 tonnes-acre and their total output came to 186,000 bottles, they decided to start making bottle-fermented sparkling wines. They engaged Clement Nowak, a Champagne-based Polish-French consultant winemaker to help and whose name at one stage actually appeared on the neck-label. For a few years Carr Taylor became the major producer – in fact almost the only producer – of bottle-fermented sparkling wines in Britain and achieved considerable success. Their *Vintage Sparkling* won a gold medal in the 1988 EWOTYC and their *Non-Vintage Sparkling* won the Jack Ward Trophy (best large volume wine) in the 1989 EWOTYC. In 1993 they won the IWSC English Wine Trophy with their *1987 Vintage Sparkling*. They also entered their wines into overseas competitions – a rarity in those days – and did surprisingly well. Their *1988 Vintage Sparkling* was awarded a gold medal at the prestigious *Concours European des Grands Vins* beating 1,800 Champagnes and other bottle-fermented sparkling wines from around the world, and in 1999, in the same competition,

64 The story was that he had been advised by the local Wine Standards inspector that he was not allowed to label it with the name of the village where the vineyard was situated, Felsted, as the village didn't have its own *appellation!*

their 1996 vintage was awarded a gold medal, this time out of 4,300 entrants. A fact that tends to get forgotten in these days of Britain's mega-vineyards planted with Champagne varieties is that the Carr Taylors were certainly the first to make serious commercial quantities of bottle-fermented sparkling wines. They did, however, only ever use what might be termed 'native' varieties for Britain: Reichensteiner, Schönburger, Kerner and Huxelrebe being the most important ones. This reliance on non-classic varieties, while it gave their wines a point of difference from other Chardonnay and Pinot based wines, also gave the wines a character more akin to Sekt or Asti than Champagne, something not all critics and commentators liked.

1985 New Hall grow grapes for the first Champagne variety sparkling wine

In 1985, Karl-Heinz Johner, then winemaker at Kenneth McAlpine's Lamberhurst Vineyards, started making sparkling wine for Piers Greenwood at New Hall Vineyards in Essex, a major grape supplier to Lamberhurst. New Hall had started growing Pinots in 1971 when they planted 0.47-ha (1.16-acres) of Ruländer, the name under which Pinot gris used to be known in southern Germany, especially Baden, and in 1973, 1974 and 1975 they continued planting and established 0.62-ha (1.53-acres) of Pinot noir in each of those years. Unknown to them at the time of planting, 50 Chardonnay vines had been included by mistake with the Pinot gris, and in 1983 (or it could have been 1984), 1,300 kgs of Pinot noir and 150 kgs of Chardonnay, 'the very last grapes picked from the harvest' according to Greenwood, were sent to Lamberhurst for Johner to make into sparkling wine. This wine was therefore almost certainly the first 'Champagne variety', bottle-fermented sparkling wine to be produced in Britain. Greenwood says the wine 'was released in December 1986 after ten months on the yeast' (which makes it more likely to have been a 1984 wine, bottled in March 1985) and is the wine which Chris Trembath, Piers' brother-in-law and now part-owner of New Hall Vineyards, remembers having at his wedding in 1987. More sparkling wine was made in 1985 (verified by a New Hall 1992 stocktaking report, which I have seen, which lists '1985 sparkling' on it). These wines were probably those that I listed as my 'Vineyard Choice' in my 1989 book *The Vineyards of England*, which said '1984/5 New Hall Sparkling Wine Medium Dry £9.95'.

McAlpine was so impressed with this wine that he asked Greenwood to plant more Pinots, exclusively for Lamberhurst, and this he did in 1988 with 3.98-ha (9.83-acres) of Pinot noir and 1.80-ha (4.45-acres) of Pinot blanc. This was at a time when most people considered Chardonnay as being too difficult to ripen, so none was planted. When I took over as winemaker at Lamberhurst in May 1988 (after Johner was dismissed for dubious blending practices) there was a stock of bottle-fermented sparkling wine in the cellars and I was told that this was made from Pinot noir, Pinot blanc and Chardonnay, all grown at New Hall. After getting Tom Stevenson to taste the wine and give it his seal of approval, it was released as *Lamberhurst Brut* in 1988 or 1989[65]. Whether this was from the same 1984/1985 grapes as New Hall's or from a subsequent vintage, I am unsure.

1986 David Cowderoy makes his first sparkling wine

The next on Britain's sparkling wine scene was David Cowderoy, son of Norman Cowderoy, who had established a vineyard at Rock Lodge near Haywards Heath in 1963 (the vineyard is now leased by Plumpton College). David had been to

65 My main memory of Tom and his wife's visit to taste the wine was that he locked his car keys in his Renault and had to wait several hours for the AA to arrive and break into it, which they did in seconds. My offers of smashing the window were soundly refused!

Wye College to study agriculture (and met Jo, now his wife who was also studying there) and after getting his degree went to Roseworthy College in Australia to study winemaking. Returning to Britain in 1986, and with the knowledge that marriage to Jo was on the horizon, he made a batch of sparkling wine using Müller-Thurgau and Reichensteiner from that vintage. Thus, when he and Jo got married in 1988, they and their friends and family were able to celebrate with what was to be the first of many batches of English sparkling wine. David continued with his experiments in sparkling wine and, working at his father's winery at Rock Lodge, produced the 1989 Rock Lodge Impresario[66] (pictured below right) which won the IWSC English Wine Trophy in 1991.

1992 Chapel Down created

In 1992 David joined forces with others to create Chapel Down Wines and one of their first wines, the non-vintage Epoch Brut, a sparkling wine, made from a blend of Müller-Thurgau, Reichensteiner and Seyval blanc, was in fact a re-badged Rock Lodge wine. The fact that Chapel Down was not using the classic Champagne varieties (which, with the exception of New Hall Vineyards, were not being grown in enough quantity for them to buy), gave them something of a marketing advantage and enabled their prices to remain reasonable – under £10 – although at the time this was at least twice that of still wines. In the end though, once Chardonnay and Pinot-based wines started to appear in 1997-98, this marketing edge disappeared and their Müller-Thurgau, Reichensteiner and Seyval blanc based wines, although very good and well-priced, were always playing second fiddle to the Champagne lookalikes in quality (and quality perception) terms, as well as in price. At much the same time, John Worontschak, winemaker at Thames Valley Vineyard (today's Stanlake Park) made a sparkling wine using Pinot noir from Ascot Vineyard, a 1-hectare (2.47-acre) vineyard planted in 1979 on Crown land near Sunninghill Park and owned by Colonel 'Robby' Robertson. Called Ascot Brut NV, it was released in 1992 and won a silver medal in the 1994 EWOTYC. Worontschak produced a number of Ascot sparkling wines from both Pinot noir and (unusually) Gamay noir, winning several silvers and bronzes between 1994 and 2004.

Mid 1980s Early days for Chardonnay, Pinot noir and Meunier based sparkling wines

The production of sparkling wines using the three classic varieties – Chardonnay, Pinot noir and Meunier – started in the mid-1980s when growers like Piers Greenwood (see above), Martin Oldaker at Surrenden Vineyard, near Ashford, (planted between 1984 and 1986) and Karen Ostborn and Alan Smalley at Throwley, near Faversham (planted in 1986)[67], both in Kent, all started growing Chardonnay and Pinot noir with the encouragement of Christopher (Kit) Lindlar. After leaving the Merrydown Wine Company[68],

66 It had originally been called 'Rock Lodge Imperial' until Moët & Chandon complained as Impérial was one of their brands.

67 The 1989 Throwley Chardonnay Sparkling won the IWSC English Wine Trophy in their 1992 competition.

68 Merrydown was partially owned by Jack Ward who was its MD, as well as being Chairman of the English Vineyards Association. Merrydown owned a small vineyard, Horam Manor, but more importantly operated a co-operative winemaking scheme for British vineyards, where in exchange for a percentage of the grapes delivered, they made your wine.

based in Horam, East Sussex, where he had been one of the winemakers since 1976, Lindlar set up as a contract winemaker, firstly at Biddenden Vineyards, and then, in 1986, at his own High Weald Winery at Grafty Green, near Ashford, Kent. Lindlar, who also supplied vines, persuaded the two Kent vineyards above to experiment with these varieties, which had until then been very unsuccessful in Britain. Brock had grown Chardonnay in his collection at Oxted but could never get it to ripen properly. Ian and Andrew Paget at Chilsdown Vineyard planted Chardonnay and also had no luck getting it to ripen. In 1981, a very dismal year for British vinegrowers, the acidity (in grams per litre) in their Chardonnay was higher than the degrees Oechsle. Ouch. Extreme unripeness was a common finding among those early growers who persevered with it, although most decided to give up and removed the offending variety. Only in really hot years would Chardonnay produce anything like ripe grapes and tolerable wine. Pinot noir, like many of the black varieties then being grown, suffered from terrible *botrytis* and was very difficult to ripen without huge losses. It is only since the arrival of better anti-botrytis sprays – initially Rovral and Ronilan, but more recently Scala, Switch and Teldor and now others – that growing fungus-sensitive varieties like Pinot noir has been possible. Meunier, in the guise of Wrotham Pinot, had always been grown in small amounts, but never used for anything other than blending with other, riper, reds. Lindlar's biggest, and subsequently best-known clients, were Stuart and Sandy Moss who decided, in 1988, to plant a vineyard at Nyetimber near Pulborough in West Sussex.

1986 Nyetimber established

The Mosses had, by all accounts, been looking at various locations to plant a vineyard – California was at one time the front runner – but it was Sandy's love of (and business in) early English oak furniture, that persuaded them that England was the place. In 1985 Hambledon Vineyards was up for sale and the Mosses viewed it and made a bid for it, but lost out to another bidder, John Patterson, who owned it until 1994. Bill Carcary, who had been at Hambledon since 1966, got to know the Mosses quite well at the time and when they then bought the 49-hectare Nyetimber estate in 1986, they asked Carcary to come and work for them as estate manager and eventually as their winemaker and got so far with this idea as to refurbish a cottage for him and offer him a contract of employment. In his discussions with them about planting a vineyard on the land at Nyetimber, Carcary remembers that it was his idea that they should plant the Champagne varieties for sparkling wine production, something he had long wanted to do at Hambledon, but which, as has been stated above, Salisbury-Jones had ruled out on cost grounds. In the end, Carcary decided for family reasons (his son had been awarded a place at the local grammar school) not to leave Hambledon and stayed, working for the new owner. Whoever actually came up with the idea of producing bottle-fermented sparkling wines on this (for the time) very large scale, the Mosses went ahead and planted the classic Champagne varieties, something which at the time was revolutionary – I said bonkers!

The vines for the Nyetimber plantings between 1988 and 1991 were sourced from the Fay nursery in France and it was to Lindlar's High Weald Winery that the first commercial vintage, the 1992, was taken for processing under the watchful eye of French winemaking consultant Jean-Manuel Jacquinot. As Lindlar modestly says, 'While they did hire Jacquinot, the winemaking buck stopped with me; that is to say, had those early vintages flopped it would definitely have been down to me.' Given the importance of the Mosses' enterprise, which when all was said and done was still something of an experiment, one has to give praise to Lindlar where it is due.

The Mosses released their first wine in 1997, the 100 per cent Chardonnay *1992 Nyetimber Première Cuvée Blanc de Blancs,* which won an

IWSC[69] Gold Medal and that competition's English Wine Trophy. Then, in 1998 they released a second wine, the *1993 Nyetimber Classic Cuvée* (a true Champagne-variety blend) which went one better, winning not only an IWSC Gold Medal and the English Wine Trophy, but also the Bottle Fermented Sparkling Wine Trophy. Suddenly, everyone woke up to the fact that good wine, even stunningly good wine, could be made from hitherto seemingly unworkable varieties – Chardonnay, Pinot noir and Meunier – and what was more, the wine could be sold at a premium price. Some of us had been wondering if things were about to change, but after these successes we realised that the game was up for German-variety based wines. With these two wines and these two significant trophies, English Sparkling Wine had definitely arrived. Nyetimber went from strength to strength, winning the Gore-Browne Trophy in 2001, 2003, 2004, 2005 and 2006 and the IWSC International Sparkling Wine Trophy in 2006, 2008 and 2009[70].

1995 Ridgeview and others join the party

A few years after the Mosses planted, another Lindlar client, Mike Roberts, also decided to establish a dedicated classic-variety, bottle-fermented sparkling wine business at Ditchling in East Sussex. Ridgeview Winery was established in 1995 with thirteen clones of Chardonnay, Pinot noir and Meunier, also supplied by the Fay nursery, and today it covers 6.48 hectares (16 acres), although it has access to grapes from a much larger area. A modern winery, with underground storage cellar, was built and equipped with the contents of the High Weald Winery, which were acquired when Lindlar closed his winery. In order to kick-start Ridgeview's production line, Chardonnay and Pinot noir grapes were bought from other growers, including Surrenden and Throwley, and the 1996 *Cuveé Merret Bloomsbury* was produced. This wine won the 2000 EWOTYC Gore-Browne Trophy, awarded for wine of the year. Since that first release, Ridgeview has produced a range of wines, all named after London squares or areas – Belgravia, Bloomsbury, Cavendish, Fitzrovia, Grosvenor, Knightsbridge and Pimlico – and the tally of awards has been impressive. They won the Gore-Browne Trophy in 2000, 2002, 2009, 2010 and 2011 and regularly win gold and silver medals in the major wine competitions. Their most notable success was probably winning the Decanter World Wine Awards International Sparkling Wine Trophy (beating four very prestigious Champagnes in the process) with their 2006 *Grosvenor Blanc de Blancs.*[71]

When first Nyetimber and later Ridgeview started selling wines and achieving the sort of prices that many in the wine business in Britain had thought impossible, the way forward for home-grown sparkling wines started to look a lot different. Following their significant commercial and competition success, plantings of the three classic Champagne varieties in Britain increased year on year and since the very warm year of 2003, several significant vineyards have been planted. The largest vineyard owner is Nyetimber, who only produce sparkling wines and only from their own vineyards. The business has changed hands twice since the Mosses owned it, and now, under its current owner, Eric Heerema, has expanded on various sites from its original 15.8 hectares to a whopping 424-hectares (1,047-acres) spread over 11 sites from Stockbridge in the west, to Chartham in the east, 177 km (110-miles) away. They have around two-thirds of their currently planted vineyards in production. Heerema told me he was always looking for more land. The next largest is Chapel Down who produce both still and sparkling wines and who own, or have planted on leased land, 341-ha (842-acres) of vineyards and

69 International Wine and Spirit Competition.

70 Since then Nyetimber are a bit more selective about which wine competitions and competitive tastings they enter.

71 Sadly, in September 2025, Ridgeview went into administration and at the time of writing, their future is unknown.

take grapes under contract from another 71-ha (175-acres). This makes a total of 412-ha (1,018-acres). Chapel Down currently (2025) have 314-ha (777-acres) in production. The third largest by vineyard size, although with not all of their planted land yet cropping, is Mark Dixon's MDCV Ltd with vineyards in East and West Sussex, Kent and Essex currently totalling 355-ha (878-acres). The next three largest sparkling wine producers are: Rathfinny with 93.5-ha; Hambledon with 90 hectares; Gusbourne with 93 hectares (although they are not currently cropping all of this); and Ridgeview with 90 hectares (both owned and under contract). Other major sparkling players (in alphabetical order) include Balfour Wine Estate, Bluebell Estates, Bolney, Camel Valley, Coates and Seely, Denbies, Evremond, Exton Park, Furleigh, Greyfriars, Hundred Hills, Laithwaites, Langham, Pinglestone, Roebuck, Simpson's, Squerryes, Tinwood, and Wiston. There are also several quite sizeable growers who mainly grow grapes under contract to the major producers. Together, there are around 25 producers who control 50 per cent of Britain's vineyard area and probably nearer 65 per cent of its sparkling wine production. And then there are the French.

2004 The French connection

Planting at Domaine Evremond in 2020

French producers, both from Champagne and elsewhere in France, have been interested in wine production in Britain for almost twenty years. In 2003, following the hottest and earliest ever harvest in Champagne, a grower and winemaker from Avizes, Didier Pierson, and his then (life) partner, British-born Imogen Whitaker, entered into a joint venture with a Hampshire farmer, and in 2004 and 2005 planted 4 hectares of the three classic Champagne varieties on a somewhat isolated, exposed and high site (up to 203 metres above sea level). The vines were planted very much *à la méthode Champenoise* with narrow rows and low trellising (an ideal height for pheasants) and farmed using the only *enjambeur* tractor in Britain. The vineyard, known as Meonhill, was subsequently bought by nearby Hambledon Vineyards following the break-up of the Pierson-Whitaker relationship. The vineyard has now been abandoned.

In around 2004–5, Champagne producer Duval-Leroy instructed the Canterbury office of Strutt & Parker, well-known land agents with strong farming links, to seek out suitable land for planting. Their brief was short, if slightly eccentric: the land must be within one hour of the Channel Tunnel, the soil must be chalky and there must have been no livestock on the land in recent years. This last request baffled most of those involved, who were told that there might be residues in the soil which would interfere with the fermentation process. (I suspect there was something lost in translation.) Although several sites were found and one selected (Squerryes Estate) and lengthy negotiations entered into, nothing came of Duval-Leroy's quest and they retreated to Épernay[72]. In 2006, on an exceptionally wet and windy June day, I took Louis Roederer boss Frédéric Rouzaud, his winemaker Jean-Baptiste Lecaillon and Mark Bingley MW, their British agent, on a tour of English vineyards simply to show them what we were up to. While this visit fuelled a lot of speculation about their intentions, it had only ever been a fact-finding trip and I never expected it to be anything else.

In 2007 Champagne producer Billecart-Salmon, in a joint venture with London wine

72 Squerryes Estate subsequently planted up and today have around 21 hectares of vineyards.

merchants Berry Brothers and Rudd, found land in Dorset for the creation of an English sparkling wine brand. They were all set to go but the financial crisis of 2008, when Champagne sales took a sudden dive, put them off the whole idea and it was shelved. When well-known wine writer and taster Steven Spurrier wanted to plant a vineyard on his wife's farm in Dorset, he first approached Duval-Leroy, who again proved difficult to negotiate with. He then he tried teaming up with the Burgundy producer Boisset whose sparkling wine house Louis Bouillot produce around 10 million bottles of Crémant de Bourgogne a year. However, Boisset were convinced that home-grown English sparkling wine couldn't sell for more than £12.99 a bottle (this was in around 2006/7) and wouldn't take things any further. Spurrier subsequently went ahead on his own, planting 10 hectares under the Bride Valley label.

2015 Domaine Evremond and Champagne Taittinger

The first Champagne house to actually take the plunge was Taittinger who, after over two years of discussions and negotiations, put their mouths and their money behind Domaine Evremond. On 26 February 2014, Patrick McGrath, a fellow Master of Wine and MD and part-owner of Hatch Mansfield Agencies Ltd (Taittinger's British agents among other things) got in touch with me saying that he wanted to discuss English sparkling wine. As a successful wine wholesaler, Hatch Mansfield had decided that they needed an English Sparkling Wine on their books. How should they go about this? Should they look at buying an existing producer? Should they go the Champagne route – buy grapes and create a blend which they would market exclusively? What were the options? Buying an existing producer I said was out. The best were not for sale and you wouldn't want any of the rest. Buy grapes? Forget it – too many of the large wineries were already chasing them and supplies were too erratic and – more importantly – too expensive (at that time). Why not go the whole hog I inquired? Buy land, plant vines, harvest grapes, make wine, bottle wine, wait three years and – hey presto – wine for sale? 'Is that a practical option? asked McGrath. That's what I help people do I said. When do we start looking for land?

The next eighteen months were spent finding a suitable site. The south-east was chosen as the target area and Kent the favourite county. Why? I knew it well, it was near the Channel Tunnel and I considered it to have the best sites and best growing conditions for vines in Britain. As Hatch Mansfield were agents for Taittinger, they had to be consulted and Pierre-Emmanuel Taittinger, grandson of the founder and most definitely an Anglophile, agreed that Taittinger would become an investor. There then followed several fraught Taittinger board meetings, at which the board eventually decided NOT to get involved. This put something of a damper on the project. However, after McGrath gave a passionate defence and promotion of the project, the board decided that they would get involved, but only if they could be majority shareholders. This was agreed and today, Taittinger and Taittinger employees hold around 55 per cent of the equity with the other 45 per cent owned by some significant minor shareholders, plus around thirty very small shareholders, mostly Hatch Mansfield employees, plus myself.

Various sites were found, looked at and rejected, but eventually what I considered to be a perfect site, south-facing, less than 100 metres above sea level, well sheltered and with mainly chalk soil was found between the villages of Selling and Chilham, eight miles south-west of Canterbury. After lengthy negotiations with the landowners, buying land they didn't want to sell and not buying blocks they wanted to sell, a deal was done and on 16 November 2015 the purchase was completed (it should have been the previous Friday but that, being the thirteenth, was rejected by the French as being inauspicious). The press launch was held on 9 December 2015 in the magnificent surroundings of Westminster Abbey, final resting place (in Poet's Corner) of Charles de Saint-Evremond. He was another Anglophile and Champagne lover who did much to introduce late-seventeenth-century Londoners to sparkling Champagne, which was becoming more popular than the predominantly still wine that was the norm at that time. The next 15 months were spent getting the site prepared, removing the apple, pear and plum trees that covered most of the site. The first 20 hectares (50 acres) of vines were planted in May 2017, another 8 hectares (20 acres) in 2019, 17 ha (42 acres) in 2020, 4-ha (10-acres) in 2021 and a further 10-ha (25 acres) in 2024, making a total of almost 60-ha (148-acres). More will be planted as land becomes available, some of the land they bought being under a short farming lease (known as an FBT) to the original owners. The first (very small) harvest was taken in October 2018, with 50+ tonnes in 2019 and then an average of 180 tonnes each year since. The first two vintages were made at Simpson's Wine Estate, but in 2020 a temporary winery was installed in a refurbished steel-framed barn on the farm and the next two vintages were made there. In June 2020, planning permission was granted for a £10 million winery and modest visitor centre and in 2022, after some covid-related delays, work started on a purpose-built semi-underground winery and cellar (costing around £11 mn) with a storage capacity for ageing 1.25 mn bottles. The first vintage was made there in 2024 and in September of the same year the first wine was released.

Following Taittinger, the floodgates hardly opened, but certainly, tongues were set wagging in Reims and Épernay and it wasn't long before the patter of French *vigneron*-sized feet were heard in England again, this time in Hampshire. Champagne Vranken-Pommery bought (from the Isaacs family, one of whom was owner of nearby Exton Park Estate, for, it is said, £20,000 an acre) a 40-hectare site near Old Arlesford and planted 19 hectares in 2018, with a further 14 hectares in 2019. In the meantime, they have collaborated with near neighbours Hattingley Valley to produce some English sparkling wines under the *Louis Pommery* brand name, the same name as they use for their Californian '*méthode Champenoise*' wine (no doubt they wish they could call it that in Britain). Boutinot, a major British-based wine wholesaler, bought Henners Vineyard in 2015 and the South African Benguela Cove Winery established vineyards in West Sussex in 2016 and 2017.

The latest large-scale overseas-based investor to set up shop in England is Mark Dixon who made his fortune with Regus, the global workspace and serviced offices company, and now owns the 121-ha Château de Berne, a Provencal Rosé producer. In around 2016 he bought Sedlescombe Organic Vineyards, followed shortly afterwards by buying Kingscote Vineyards, East Grinstead (which has now been sold). Following the purchase of these, new vines were planted at both places, plus at Bodiam Vineyard, which Sedlescombe had leased since 1994. Further land was leased from the owners of the Bodiam vineyard, the Sternberg family, at Old Place Farm, Sandhurst, just below Sandhurst Church. Here, around 40 ha (100 acres) of vines were planted in 2019, somewhat perversely (and baffling to local vineyard owners) at an angle across a perfectly fine south-facing slope. Not content with this number and area of vineyards, the already-organic Court Lodge and Brookers farms in the hamlet of Luddesdown near Gravesend were purchased and in 2019 some of the land was planted with 177-ha of Champagne varieties. More land has subsequently been planted here with the planted total now 202-ha (500-acres). In 2020 another new vineyard, this time a mere 38-ha in Essex in the village of Althorne facing onto the River Crouch in the by now much-bevined CM3 postcode. In total, Dixon, trading as MDCV or Silverhand Estate, now has control over 283-ha (700-acres) of vineyards, most of which is either organic, biodynamic or in conversion. Quite what the long-term plan is for all these vineyards is not clear. Both Sedlescombe and Kingscote were put on the market in early

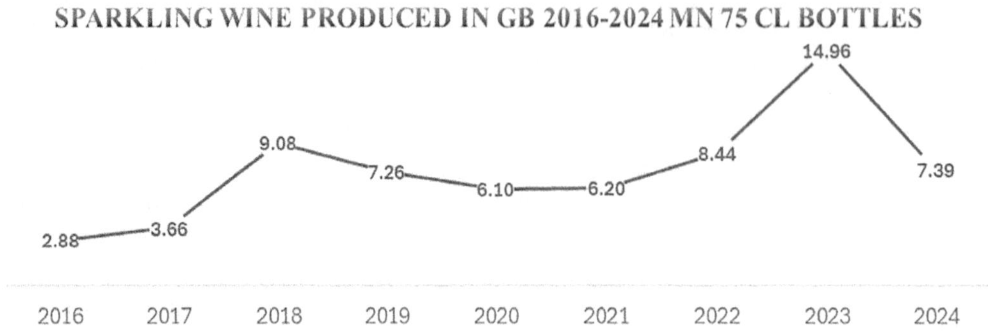

2020 as they 'didn't fit in with the business plan going forward' and Kingscote was eventually sold in 2025. They are currently offering their Charmat method sparkling wine at £13.60 a bottle. It will be interesting to see how this empire of wine develops.

Sparkling wine expands

Other southern French, Italian and even Californian producers have been on reconnaissance missions to look at vineyards in Britain and although they are all surprised at how cheap vineyard land is compared to where they come from, their main discovery is how low yields are. With still wines selling in the £10–15 range and sparkling wines retailing between £16 and £35, can you really make money when you only harvest 5 tonnes-ha (2 tonnes-acre)? Those yields might be okay if you are selling high-priced Champagne, Bordeaux or Burgundy, or New World wines with more than 95 Parker points, but they do not work for relatively modestly priced wines. The latest and perhaps the most important overseas entrant to the GB wine market is Jackson Family Wines, the Californian-based wine company who own or have substantial interests in over 40 wineries and and 4,452-ha (11,000-acres) of vines. After several years' research they have now invested in GB, buying an extremely expensive plot of land in the Crouch Valley and planting 14-ha (35-acres) of Pinot and Chardonnay in 2024. In 2025 they leased land from Wiston Estate in West Sussex and planted a further 9-ha (22-acres) of vines. They also bought finished wine to make a sparkling wine cuvée which is now ageing, and in June 2025 released a 2024 still Chardonnay called 'Marbury'. With their level of family commitment and world-wide marketing expertise, how they market their wines will be interesting to see.

Sadly, the rapid expansion of vineyards and wine production has led to several casualties. In 2023, Hambledon Vineyards found itself in financial straits and was purchased in its entirety by two industry majors, Berry, Bros & Rudd and Symington Family Estates. In late 2024, Highweald Wines had financial problems and went into administration in April 2025. They have since been refinanced and re-emerged under new management. In 2025 Gusbourne, which was a PLC, was removed from the AIM stock market by its majority shareholder and backer, Lord Michael Ashcroft, and took the decision not to take any crop from the 32-ha of vines it has in West Sussex citing excess stocks as the reason. It is still trading from its Kent site where it has 61-ha of vines. The final casualty, but by no means the least, is Ridgeview which many considered to be one of the best run and best managed wine producers in the country. In October 2025, after reporting a £1.5 mn loss for 2023, they went in administration. They are still trading and hoping to find a buyer for the business.

What's the future for English sparkling wine?

The future for English sparkling wines is as uncertain as most things these days. There is a current imbalance between demand and supply with many producers holding large stocks (some say very large) in relation to their current sales. Plus, around 50 per cent of the sparkling wine vineyards currently planted do not yet have wine for sale, either because they have not yet started to crop, or because the wine they have produced is still maturing and not yet ready for sale. The production of sparkling wine in the last three years (2022-24) has averaged 10 mn bottles a year, whereas sales have averaged 6 mn bottles a year (2021-23). With 300-350 ha of new plantings coming on-stream every year, and with vineyard plantings showing little sign of stopping, at some point, demand and supply have to reach an equilibrium. Many think that substantial price reductions of both wine and grapes cannot be far away.

Appendix II
The story of Wrotham Pinot

The vine that Edward Hyams discovered in Wrotham in the early 1950s was in all probability the variety known as 'Miller's Burgundy' which had been widely grown on walls and in gardens in Great Britain for at least two hundred years. Philip Miller's *Gardener's Dictionary*, first published in 1735, has an entry for 'the black Cluster or Munier [sic] grape, as it is called by the French, from the hoary down of the leaves in summer --- which some call the Burgundy grape'. Archibald Barron writing in his book, *Vines and Vine Culture,* the standard Victorian work on grape growing and first published in 1883, states that the variety was: *found by* [the famous horticulturalist] *Sir Joseph Banks in the remains of an ancient vineyard at Tortworth, Gloucestershire* – a county well known for its medieval vineyards. Hyams took cuttings to Brock at Oxted who said that when compared to supplies of Meunier from France, the variety now known as Wrotham Pinot: *had a higher natural sugar content and ripened two weeks earlier*. Hyams, ever the journalist in search of a good story, claimed that this vine had been left behind by the Romans although he provided no evidence for this.

Brock, in his *Report No. 3: Progress With Vines and* Wines of 1961, states that 'Wrotham Pinot appears to be better adapted to this climate than the standard Pinot grapes' and 'is a consistent cropper and ripens here in good years in the open vineyard'. i.e. Not under cloches. He continues, 'It would probably be suitable for a really highly flavoured Burgundy type of red wine in favoured sites in the south of England, but it is not suitable where the conditions are not very good'. Brock sold cuttings and the variety became quite popular in early vineyards, although it is unlikely that any vines from the cuttings supplied by Brock survive in any of today's UK vineyards. Jack Ward, writing in his 1984 book *Vine Growing in the British Isles*, says that Wrotham Pinot 'was given a fair trial in the Beaulieu vineyard and contributed to the production of a rosé wine' but that 'its performance was adequate but hardly good enough, for it was later grubbed in favour of other varieties'. Gillian Pearkes, writing in the 1989 version of her 1982 book *Vinegrowing in Britain* says that Meunier 'suffers from the same inadequacies' as Pinot noir which she states 'is seldom if ever even a mediocre wine; it is thin, pale and lacking in intensity of bouquet and depth of flavour'. Despite the fact that today all plantings of Meunier in the UK stem from French and German nurseries, the name Wrotham Pinot is still a legally acceptable synonym for this variety, although the name is not used at all in the UK.

In 2004, a well-known grower from California, Richard Grant Peterson, who was winemaker at Beaulieu Vineyards in Napa after the famous André Tchelistcheff, announced that he had a vineyard in Yountville in the Napa Valley, planted with Wrotham Pinot vines propagated from cuttings taken from the original Wrotham Pinot vine in 1980. At the time, Peterson was a judge at the International Wine and Spirits Competition (IWSC) and had been told about a vine growing wild against a stone wall in the village of Wrotham. He then claims to have inspected it and found that it had tiny white hairs on the upper surface of the leaves and was unlike any vine he had ever seen. In addition, this vine: 'had developed a considerable immunity to Powdery Mildew'. On his website, Peterson states that the original vine in Wrotham died in 1990, but that: *there are now at least two new wild seedlings growing a few feet from where the mother vine had stood.* This statement would appear to show that Peterson knows where the original vine was. Despite having been in contact with Peterson, I have been unable to verify whether he personally visited the cottage in Wrotham or who the 'local winemakers' were that had made the wine he tasted. Ignoring the fact that a 'seedling' cannot possibly be the same variety as the vine variety that produced the bunch of grapes from which it came, I find the whole story difficult to believe. Peterson also states that Dudley Quirk, who had a vineyard at Chiddingstone in Kent in the 1980s and 1990s, grew Wrotham Pinot. Again, this does not ring true to me. I knew Quirk (and his wife and his children) quite well and one thing is for certain, he liked making up tall tales. As the great newspaper proprietor William Randolph Hearst said 'Don't let the truth get in the way of a good story'.

On his website, Peterson states 'English viticulture scholars eventually came to the conclusion that the Wrotham vine was a natural seedling of Pinot noir vines that the early Romans brought to England 2000 years ago' without any proof whatsoever that the Romans even had proper vineyards in Britain. Ignoring the 'seedling' question again, quite who these 'English viticulture scholars' were is not known, but possibly in reference to Ray Brock at Oxted, although he never mentioned this to me when I interviewed him (several times) for my UK Vineyards Guide before he died in 1999.

Peterson also states that he tasted a sparkling *Blanc de Noirs* wine from the variety made by 'local winemakers' and immediately recognised its potential. He imported cuttings into California and, 'after a quarantine of many years as required by law', vines were propagated and a 0.8-ha (2-acre) vineyard established which he grows without the need to spray them with sulphur against Powdery Mildew. Peterson had a vine analysed by the University of California's Davis wine department who pronounced the vine's DNA to be 'identical to that of Pinot noir'.

Peterson made a bottle-fermented rosé sparkling, which sold for $60 a bottle and its quality is obviously excellent as the 2001 vintage won a Double Gold at the US National Women's Wine Competition. On the website www.princeofpinot.com it says that he now produces a sparkling wine and a still wine in alternate years 'because of the small number of vines available'. WineSearcher list a $60 *Richard Grant North Coast Blanc de Noir* which doesn't mention Wrotham Pinot, although the label shown does, and says it's a '2014 Reserve' from the Oak Knoll District which is at the southern end of the Napa Valley. WineSearcher also says that the still Pinot noir was 'last available in November 2017'.

The Wrotham Historical Society published a booklet called *Farming in Wrotham Through the Ages* which I purchased for £3 in the hope that it might provide me with some useful information. The booklet mentions that: *there is reputed to have been a Roman vineyard on the slopes of the North Downs above Wrotham* (funny how reports about Roman vineyards always start with 'reputed' or 'thought to have been') but about the location of the cottage it gives little help. The booklet merely says that the cottage wall was *on the main road in the village* which is hardly giving the game away. The text in the booklet is more or less the same as can be found on Peterson's website: www.richardgrantwine.com. Amuse Bouche Wines of Napa City sell a wine labelled 'Richard G. Peterson Pinot Noir from the Santa Lucia Highlands AVA' claiming that 'in 1980, Dick Peterson brought the first and only Wrotham cuttings back to California and planted them in the Santa Lucia Highlands AVA just south of Monterey Bay'.

Today in Britain there are to my knowledge no plantings of Wrotham Pinot that stand up to scrutiny and the only vineyard I know of that claimed to have any was Barnsole Vineyard near Canterbury, Kent. The owner told me that he got his vines from Sunnybank Vine Nursery which state on their list of vines for sale that '*DNA analysis in 2019 showed this to match Pinot Meunier – mildew prone*'. Given that this clone is 'mildew prone' it hardly matches up to Peterson's clone which he says is completely mildew resistant. The Wine Standards vine variety database, the official register of vines planted in GB, does not list any Wrotham Pinot.

A good few years ago I was introduced to a

producer in Novia Scotia called Lightfoot and Wolfville who claimed to have some (he wouldn't tell me where he got the vines from) and when he realised that I knew something about the supposed variety, he changed the subject. On their website now I note that they say that the vines in their 'mystery Pinot block' were 'sourced from cuttings believed to be a unique, early ripening, powdery mildew-resistant clone of Pinot noir' which 'Richard Grant Peterson brought to North America'.

Its an interesting story and one that I guess will linger on for years given that fact that the internet never forgets.

Appendix III

Distribution of vineyards in Great Britain

Vineyards in Great Britain - 2024	
County	**Hectares**
Kent	1,238.86
West Sussex	670.83
Essex	468.56
East Sussex	460.60
Hampshire	426.10
Surrey	129.63
Dorset	97.07
Devon	87.11
Gloucestershire	77.88
Suffolk	76.41
Berkshire	68.10
Norfolk	65.44
Herefordshire	65.20
Cornwall	56.20
Oxfordshire	51.64
Wiltshire	47.08
Hertfordshire	39.23
Worcestershire	37.98
Shropshire	37.04
Somerset	30.27
Another 25 counties	258.55
Wales	91.35
Total GB	**4,581.13**

Source: Wine Team at the Food Standards Agency 2024

Table 38

Vineyards in Great Britain - 2024		
Region	**Hectares**	**Percentage**
East Anglia	673.01	14.69%
South East	2,510.64	54.80%
Wessex	594.21	12.97%
West	377.90	8.25%
T&CVA	138.20	3.02%
Midlands & North	195.82	4.27%
Wales	91.35	1.99%
Total	**4,581.13**	**100.00%**

Source: Wine Team at the Food Standards Agency 2024

Table 39

Appendix IV
Wye College Experimental Vineyard

In November 1971, Wye College, the University of London's agricultural department near Ashford in Kent, which specialised in research into hops, had suggested that it might undertake some research into grapevines by planting a trial vineyard. The EVA launched an appeal and the sum of £10,000 was raised towards the costs, with donations coming from members and interested organisations. Vines were donated by various German, Swiss and Austrian universities, together with some from Weingut Louis Guntrum in Nierstein. In total, 850 vines of 17 different varieties were obtained (not all of them on the permitted lists) and planted in 1973, with more to follow in 1974, 1975, 1980 and 1981. The vineyard eventually covered 1.3 acres (0.526 ha).

The vines were trained on three different trellising systems: Double Guyot, Lenz Moser and GDC. Bob Farrar, then in charge of the management of the College's hop gardens, took great personal interest in the vineyard, and was really responsible for seeing that it was maintained at minimal expense to the EVA, making sure that it was sprayed with fungicides left over from spraying the hop gardens. The yield and performance statistics gathered gave valuable information about the different varieties being grown there. The grapes harvested were either sold to Merrydown or sent to Long Ashton Research Station in Bristol to be made into wine. Eventually, the vineyard site was required by Wye College for a new hop garden and in 1984 the vines were grubbed-up. Although it only lasted just over a decade, the Wye College vineyard introduced Regner, Würzer, Bacchus, Faberrebe and Zweigeltrebe to GB vineyards.

www.ingramcontent.com/pod-product-compliance
Lightning Source LLC
LaVergne TN
LVHW082349200926
848481LV00012B/3511
* 9 7 8 1 9 1 6 3 2 9 6 5 2 *